64 Advances in Polymer Science
Fortschritte der Hochpolymeren-Forschung

Polymer Membranes

Editor: M. Gordon

With Contributions by
H. Bader, K. Dorn, K. Hashimoto, B. Hupfer,
J. H. Petropoulos, H. Ringsdorf, H. Sumitomo

With 78 Figures and 15 Tables

Springer-Verlag
Berlin Heidelberg GmbH
1985

ISBN 978-3-662-15215-7   ISBN 978-3-540-38960-6 (eBook)
DOI 10.1007/978-3-540-38960-6

Library of Congress Catalog Card Number 61-642

© Springer-Verlag Berlin Heidelberg 1985
Originally published by Springer-Verlag Berlin Heidelberg New York in 1985
Softcover reprint of the hardcover 1st edition 1985

2154/3020–543210

Editors

Editorial

With the publication of Vol. 51, the editors and the publisher would like to take this opportunity to thank authors and readers for their collaboration and their efforts to meet the scientific requirements of this series. We appreciate our authors concern for the progress of Polymer Science and we also welcome the advice and critical comments of our readers.

With the publication of Vol. 51 we should also like to refer to editorial policy: *this series publishes invited, critical review articles of new developments in all areas of Polymer Science in English (authors may naturally also include works of their own).* The responsible editor, that means the editor who has invited the article, discusses the scope of the review with the author on the basis of a tentative outline which the author is asked to provide. Author and editor are responsible for the scientific quality of the contribution; the editor's name appears at the end of it.

Manuscripts must be submitted, in content, language and form satisfactory, to Springer-Verlag. Figures and formulas should be reproducible. To meet readers' wishes, the publisher adds to each volume a "volume index" which approximately characterizes the content.

Editors and publisher make all efforts to publish the manuscripts as rapidly as possible, i.e., at the maximum, six months after the submission of an accepted paper. This means that contributions from diverse areas of Polymer Science must occasionally be united in one volume. In such cases a "volume index" cannot meet all expectations, but will nevertheless provide more information than a mere volume number.

From Vol. 51 on, each volume contains a subject index.

Editors   Publisher

Table of Contents

Polymeric Monolayers and Liposomes as Models for Biomembranes

How to Bridge the Gap Between Polymer Science and Membrane Biology?

Hubert Bader, Klaus Dorn, Bernd Hupfer, Helmut Ringsdorf
Institut für Organische Chemie, Johannes Gutenberg-Universität J.-J.-Becher-Weg
18–20, D-6500 Mainz, FRG

Polymer chemists as poachers in foreign grounds? Why not? Macromolecular chemistry has become a mature science with all advantages and handicaps of a well-established scientific discipline: many heights are conquerred and the harvest is abundant, but adventures and the future might be elsewhere. Besides, in these times of bottomed out industrial profits in common plastics, future polymer chemistry cannot be limited to repetitive improvement of already successful mass polymers but should rather expand into neighboring fields of material science as well as life science where "polymer thinking" might help to overcome difficulties. — First hesitant steps on the bridge towards membrane biology have been made.

This contribution will first deal with stabilization of membrane model systems (monolayers, black lipid membranes, vesicles) in general. The desired further biological functionalization of these stabilized polymeric membranes is possible via incorporation of proteins. Addition of natural lipids to polymerizable membranes and enzymatic degradation of the unpolymerized component after polymerization allows selective opening of otherwise stable compartments. Cell-cell recognition — of vital importance in immunological reactions — can be mimicked with polymeric vesicles carrying sugar headgroups. Finally, attempts are made to unite biological specification of natural cells and toughness of polymerized membranes via cell-vesicle fusion.

Macromolecular chemistry is far from being able to provide ultimate solutions for specific problems in membrane biology or life science in general. However in "going for it" polymer chemists can only benefit from the many chances to learn. The key attitudes to enjoy this adventure are courage to enter neighboring fields and, especially, the willingness to cooperate. This is perfectly illustrated in Fig. 1 by H. von Saalfeld and his artistic view of scientists and their unpretended willingness to cooperate [90].

Fig. 1. "Well, dear friend and colleague, I really appreciate your personal attitude and engagement for our profound area of research and I am looking forward to a close and fruitful cooperation"

1 Mother Nature Forms Stable Membranes. Can Chemists Reach her Standards?

Membranes of plant and animal cells are typically composed of 40–50% lipids and 50–60% proteins. There are wide variations in the types of lipids and proteins as well as in their ratios. Arrangements of lipids and proteins in membranes are best considered in terms of the fluid-mosaic model, proposed by Singer and Nicolson [1]. According to this model, the matrix of the membrane (a lipid bilayer composed of phospholipids and glycolipids) incorporates proteins, either on the surface or in the interior, and acts as permeability barrier (Fig. 2). Furthermore, other cellular functions such as recognition, fusion, endocytosis, intercellular interaction, transport, and osmosis are all membrane mediated processes.

Much of our chemical understanding of membrane structures has been obtained through the investigation of models, most of which are based solely on the reconstruction of the lipid bilayer part of biomembranes. In contrast to biomembranes, rebuilt bilayers such as lipid vesicles are not stable in the long term. Obviously, nature finds additional means for creating membranes of high stability:

1. Besides the so-called 'integral' membrane proteins (which are embedded in the hydrophobic part of the bilayer) there are peripheral proteins adsorbed on the hydrophilic surface of the membrane. Some of these peripheral proteins act as support, because they are associated with several integral proteins. A well known example is spectrin, situated at the inside of the erythrocyte membrane [2].

2. The so-called 'coated vesicles' are an example of an enveloped membrane [3]. A phospholipid vesicle within a cell is coated by a polypeptide and resembles a football in a net. A comparable feature of membrane coating is found in cell walls of bacteria [4]. These cell walls consist of polysaccharides which are cross-linked by oligopeptides. It is remarkable that this extreme stabilization by an exogenous support is found in bacteria. As parasites in foreign tissues, they have to be especially resistant.

How can a stabilization of biomembranes be achieved synthetically? An attempt to mimic a support similar to the spectrins seems unfeasible, for very little is known as yet about the interactions between peripheral and integral proteins. An increase of stability via polymer coatings, as in the case of bacteria membranes, sounds more realistic and is, in fact, used for immobilizing living cells. This coating however, prevents cell-cell contact and hence interaction of different cells [5].

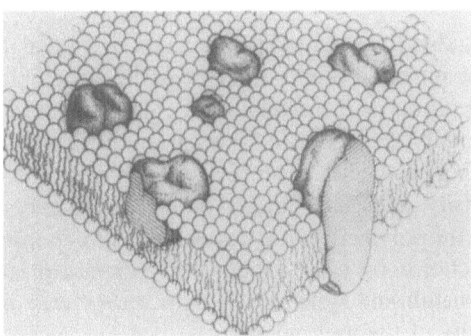

Fig. 2. The lipid-globular protein mosaic model with a lipid matrix (fluid mosaic model [1]); schematic three dimensional and cross-sectional view. The solid bodies with stippled surfaces represent the globular integral proteins, which are randomly distributed in the plane of the membrane

The most convenient and realistic approach for stabilizing model membrane systems seems to be a method using membrane lipids for stabilization. A typical experiment, although used for a different purpose, has been described by Khorana et al. [6]. These authors incorporated lipids carrying photoreactive groups into model membranes and could covalently link the lipid molecules by UV irradiation. This concept has been generalized to a great extent by making use of polymerization reactions. An extensive compilation of amphiphilic molecules carrying polymerizable groups, methods used for their polymerization, and their characterization in model membranes is given in the following chapters.

2 Polymerizable Lipids

The most common lipid components of cell membranes are phospholipids. The structures of typical representatives of these amphiphilic molecules (with hydrophobic alkyl chains and hydrophilic head group) are illustrated in Fig. 3.

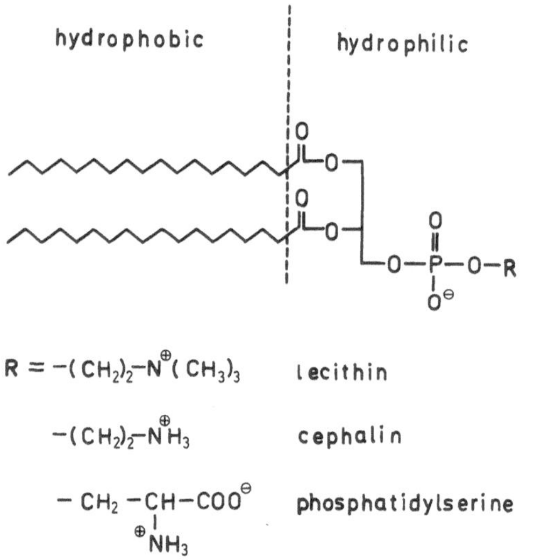

$R = -(CH_2)_2-N^{\oplus}(CH_3)_3$ lecithin

$-(CH_2)_2-N^{\oplus}H_3$ cephalin

$-CH_2-CH-COO^{\ominus}$ phosphatidylserine
         $|$
         $^{\oplus}NH_3$

Fig. 3. Structures of the most common biomembrane phospholipids

If one intends to synthesize polymerizable lipids to build up membranes of high stability the polymerizable group may be introduced into different parts of the lipid molecule, i.e., into the hydrophilic head group or the alkyl chain (Fig. 4) [7].

All four systems illustrated in Fig. 4 exhibit properties differing from those of cell membranes. Methods a–c have no influence on the head groups and preserve physical properties, such as charge, charge density, etc. The fluidity of the hydrocarbon core, however, is drastically decreased by the polymerization process. In case d, fluidity is not affected, but there is no free choice of head groups. In comparison to biomembranes, all polymerized model membrane systems will show an increase in viscosity and a decrease in lateral mobility of the molecules.

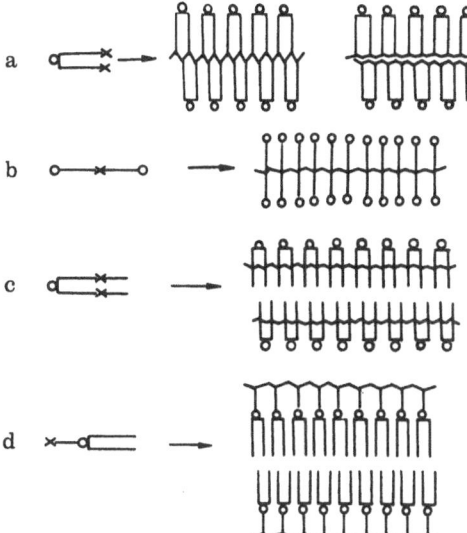

Fig. 4a–d. Possible ways to synthesize polymeric model membranes [7] (X — polymerizable group). a — c: Polymerization with preservation of head group properties; d: polymerization with preservation of chain mobility. For examples of corresponding monomers, see Table 1

All four types of polymerizable lipids shown in Fig. 4 have been realized synthetically. In this context, "one need not attempt to reproduce mother nature slavishly" (Fendler [8]). Kunitake [9] was able to show that simple molecules like dialkyldimethylammonium salts also form bilayer assemblies. Fuhrhop [10] and Kunitake [11] could accomplish the formation of monolayer liposomes with molecules containing only one alkyl chain and two hydrophilic head groups. Acryloylic and methacryloylic groups (type a and d, Table 1), as well as diacetylenic, butadienic, vinylic and maleic acid groups (type b and c), have been used as polymerizable moieties. A compilation of amphiphilic, photopolymerizable molecules is given in Table 1.

Besides polymerization, another type of polyreaction can be used for stabilizing model membrane systems. Recently, Fukuda et al. [28] described polyamide formation via polycondensation in monolayers at the gas/water interface (definition of monolayers see Sect. 3.2). Long-chain esters of glycine and alanine were polycondensed to yield *non-oriented* polyamide films of polyglycine and polyalanine.

The self condensation of esters of long-chain α-amino acids (methyl 2-aminooctadecanoate (*39*), docosanyl 2-aminooctadecanoate (*40*), methyl 2-aminohexacosanoate (*41*), docosanyl 2-aminohexacosanoate (*42*) is used to prepare *oriented* polypeptide films, which are models for biological membranes. [52]

$$H_3C-(CH_2)_{15}-\underset{\underset{NH_2}{|}}{CH}-CO-O-CH_3 \qquad H_3C-(CH_2)_{15}-\underset{\underset{NH_2}{|}}{CH}-CO-O-(CH_2)_{21}-CH_3$$

$$(39) \qquad\qquad\qquad (40)$$

$$H_3C-(CH_2)_{23}-\underset{\underset{NH_2}{|}}{CH}-CO-O-CH_3 \qquad H_3C-(CH_2)_{23}-\underset{\underset{NH_2}{|}}{CH}-CO-O-(CH_2)_{21}-CH_3$$

$$(41) \qquad\qquad\qquad (42)$$

H. Bader, K. Dorn, B. Hupfer, H. Ringsdorf

Table 1. Polymerizable Lipids and Surfactants

Type	Surfactant		Ref.

a)

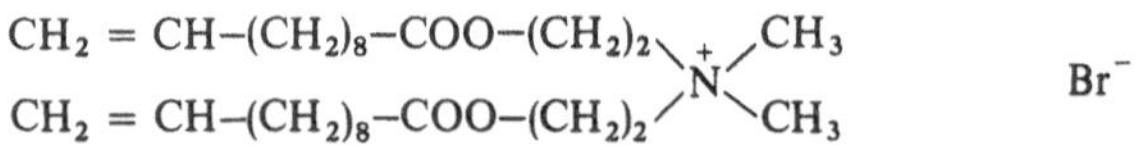

$$CH_2 = CH-(CH_2)_8-COO-(CH_2)_2 \diagdown \overset{+}{N} \diagup {}^{CH_3} $$
$$CH_2 = CH-(CH_2)_8-COO-(CH_2)_2 \diagup \quad \diagdown {}^{CH_3} $$

Br⁻   *1*   12)

$$\overset{CH_3}{\underset{|}{CH_2}} = C-CO-NH-(CH_2)_{10}-COO-(CH_2)_2 \diagdown \overset{+}{N} \diagup {}^{CH_3}$$
$$CH_2 = \underset{|}{C}-CO-NH-(CH_2)_{10}-COO-(CH_2)_2 \diagup \quad \diagdown {}^{CH_3}$$
$$CH_3$$

Br⁻   2   7)

$$\overset{CH_3}{\underset{|}{CH_2}} = C-COO-(CH_2)_{11}-COO-CH_2$$
$$CH_2 = \underset{|}{C}-COO-(CH_2)_{11}-COO-CH$$
$$CH_3 \qquad\qquad CH_2-O-\overset{O}{\overset{\|}{P}}-O(CH_2)_2-\overset{+}{N}(CH_3)_3$$
$$\underset{O_-}{}$$

3   13)

$$H_3C-(CH_2)_{15} \diagdown \overset{+}{N} \diagup {}^{CH_3}$$
$$CH_2{=}C-COO-(CH_2)_{11} \diagup \quad \diagdown {}^{CH_3}$$
$$\underset{CH_3}{}$$

Br⁻   4   14)

$$H_3C-(CH_2)_{17} \diagdown \overset{+}{N} \diagup {}^{CH_3}$$
$$CH_2{=}C-COO-(CH_2)_{11}-COO-(CH_2)_6 \diagup \quad \diagdown {}^{CH_3}$$
$$\underset{CH_3}{}$$

Br⁻   5   15)

$$H_3C-(CH_2)_{14}-COO-CH_2$$
$$H_2C = C-COO-(CH_2)_{11}-COO-CH$$
$$CH_3 \qquad\qquad CH_2-O-\overset{O}{\overset{\|}{P}}-O(CH_2)_2-\overset{+}{N}(CH_3)$$
$$\underset{O_-}{}$$

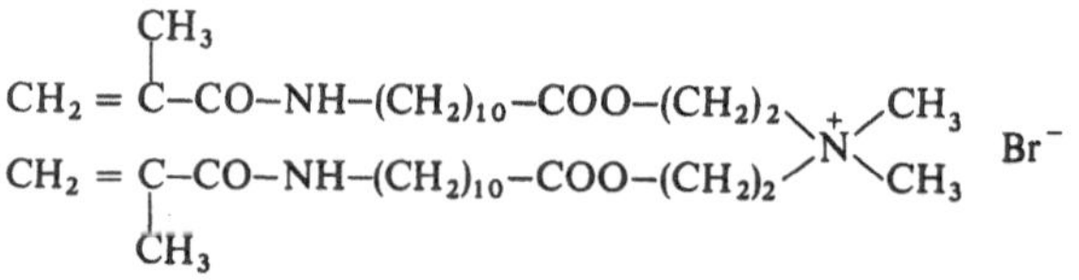

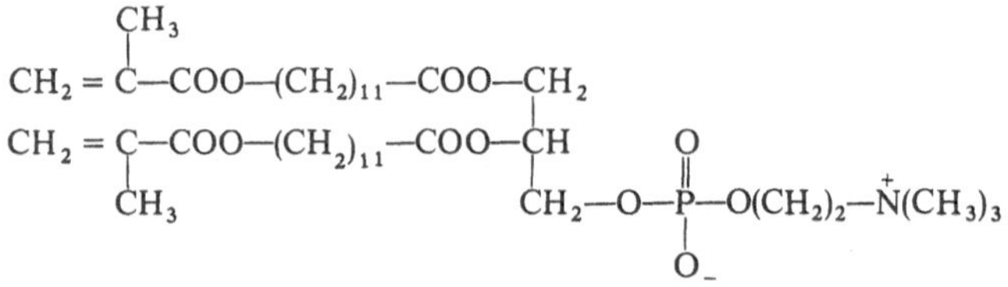

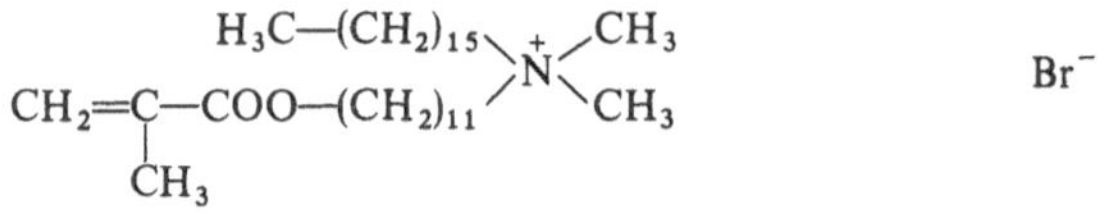

6   13)

$$\overset{CH_3}{\underset{|}{H_2C}} = C-COO-(CH_2)_{11}-COO-CH_2$$
$$H_3C-(CH_2)_{14}-COO-CH$$
$$CH_2-O-\overset{O}{\overset{\|}{P}}-O(CH_2)_2-\overset{+}{N}(CH_3)_3$$
$$\underset{O_-}{}$$

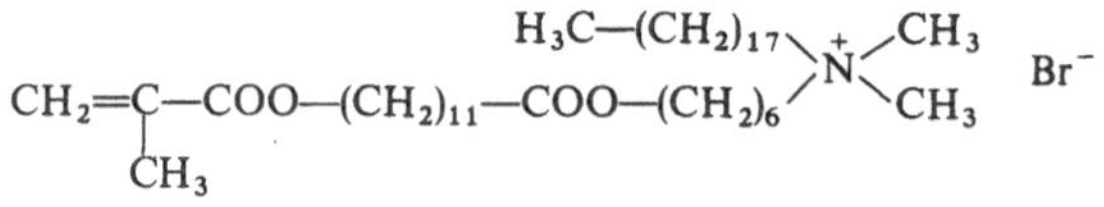

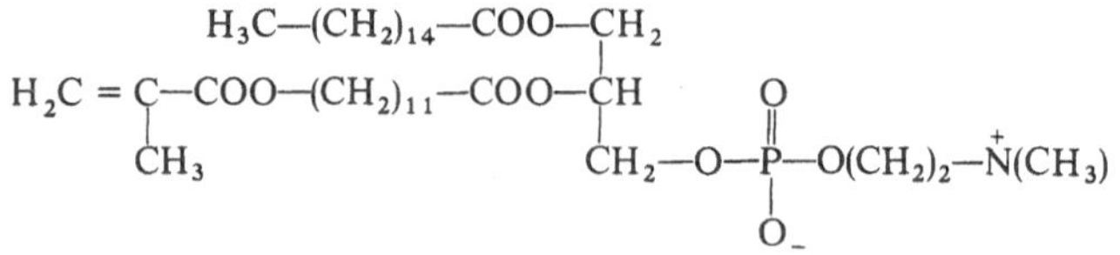

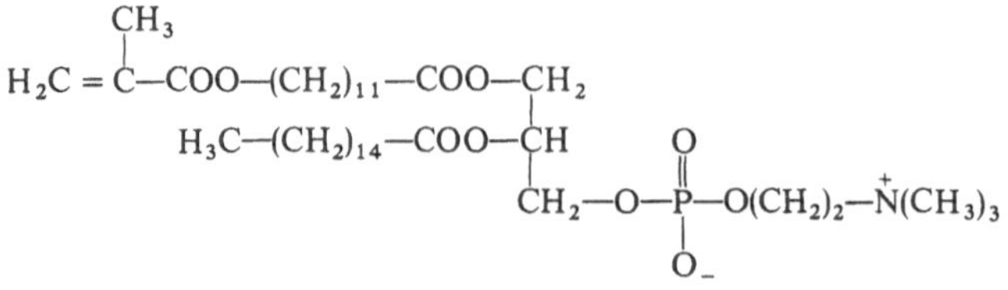

7   13)

Table 1. (continued)

Type	Surfactant		Ref.

a)

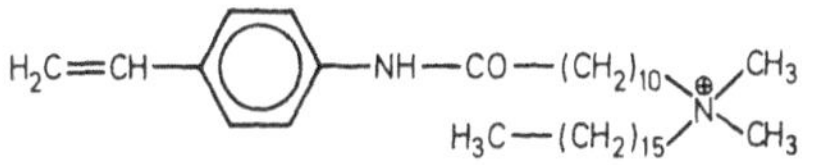

$H_2C=CH-\langle\bigcirc\rangle-NH-CO-(CH_2)_{10}-\overset{\oplus}{N}(CH_3)(CH_2)_{15}CH_3$, $Br^{\ominus}$ 8   12)

$H_2C=CH-(CH_2)_8-COO-(CH_2)_2$, $H_2C=CH-(CH_2)_8-COO-(CH_2)_2$ $N-\overset{O}{\overset{\|}{P}}(OH)_2$ 9   12)

$H_2C=CH-(CH_2)_8-COO-(CH_2)_2$, $H_2C=CH-(CH_2)_8-COO-(CH_2)_2$ $\overset{\oplus}{N}(CH_3)(CH_2)_2-OH$, $Br^{\ominus}$ 10   12)

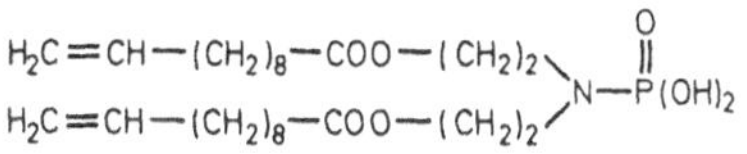

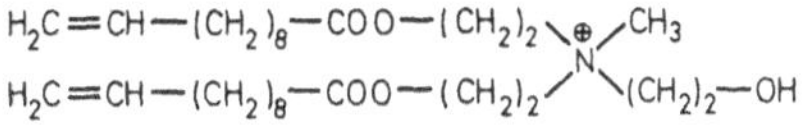

$H_2C=CH-(CH_2)_8-COO-(CH_2)_2$, $H_2C=CH-(CH_2)_8-COO-(CH_2)_2$ $\overset{\oplus}{N}(H)(CH_2)_2-SO_3^{\ominus}$ 11   12)

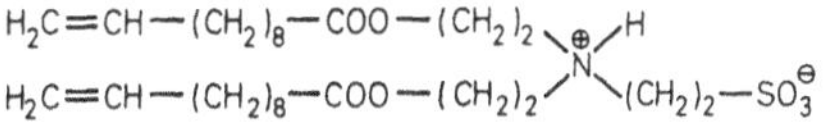

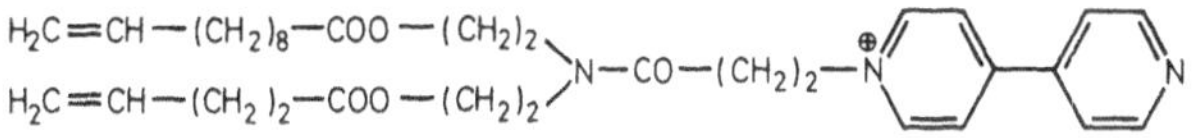

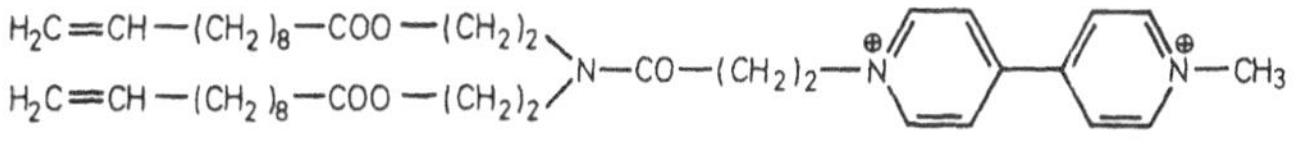

$H_2C=CH-(CH_2)_8-COO-(CH_2)_2$, $H_2C=CH-(CH_2)_2-COO-(CH_2)_2$ $N-CO-(CH_2)_2-\overset{\oplus}{N}\langle\text{bipyridine}\rangle N$, $Br^{\ominus}$ 12   16)

$H_2C=CH-(CH_2)_8-COO-(CH_2)_2$, $H_2C=CH-(CH_2)_8-COO-(CH_2)_2$ $N-CO-(CH_2)_2-\overset{\oplus}{N}\langle\text{bipyridine}\rangle\overset{\oplus}{N}-CH_3$, $Br^{\ominus}, J^{\ominus}$ 13   16)

$H_2C=CH-(CH_2)_9-O$, $H_2C=CH-(CH_2)_9-O$ $\overset{O}{\overset{\nearrow}{P}}\overset{\diagdown}{OH}$ 14   17)

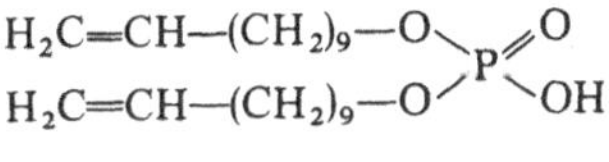

b) $HOOC-(CH_2)_8-C\equiv C-C\equiv C-(CH_2)_8-COOH$ 15   18)

$HO-(CH_2)_9-C\equiv C-C\equiv C-(CH_2)_9-OH$ 16   18)

$H_2O_3P-O-(CH_2)_9-C\equiv C-C\equiv C-(CH_2)_9-O-PO_3H_2$ 17   18)

c) $H_3C-(CH_2)_n-C\equiv C-C\equiv C-(CH_2)_8-COO-CH_2$
$H_3C-(CH_2)_n-C\equiv C-C\equiv C-(CH_2)_8-COO-CH$
$\qquad\qquad\qquad\qquad CH_2-O-\overset{O}{\overset{\|}{P}}-O-(CH_2)_2-\overset{+}{N}(CH_3)_3$
$n = 9, 11, 12$ $\qquad\qquad\qquad\qquad O_-$ 19, 20, 21)

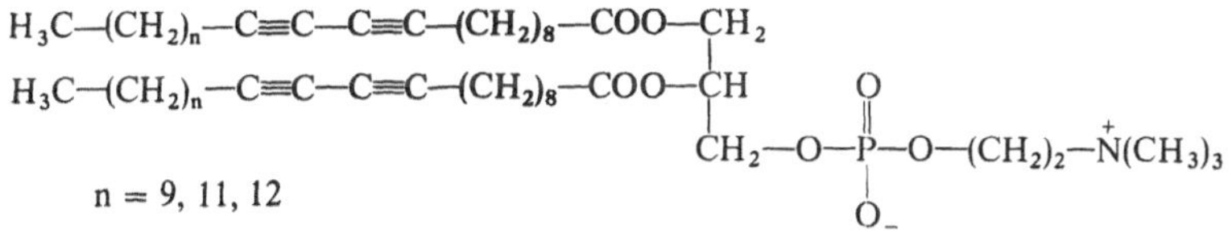

18

Table 1. (continued)

Type	Surfactant		Ref.

c) $H_3C-(CH_2)_{12}-CH=CH-CH=CH-COO-CH_2$
      $H_3C-(CH_2)_{12}-CH=CH-CH=CH-COO-CH$

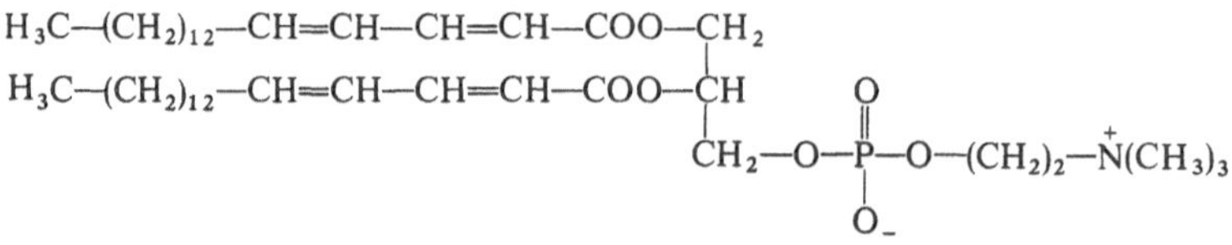

$$CH_2-O-\overset{O}{\underset{O_-}{\overset{\|}{P}}}-O-(CH_2)_2-\overset{+}{N}(CH_3)_3$$

                                                        19   19, 22)

$H_3C-(CH_2)_n-C\equiv C-C\equiv C-(CH_2)_8-COO-(CH_2)_2$
$H_3C-(CH_2)_n-C\equiv C-C\equiv C-(CH_2)_8-COO-(CH_2)_2$ $\overset{+}{N}\overset{CH_3}{\underset{CH_3}{}}$ Br^- *20*   19, 21)
                                                                          $n = 9, 12$

$H_3C-(CH_2)_9-C\equiv C-C\equiv C-(CH_2)_9-O$
$H_3C-(CH_2)_9-C\equiv C-C\equiv C-(CH_2)_9-O$ $\overset{O}{\underset{OH}{P}}$ *21*   21)

$H_3C-(CH_2)_{12}-C\equiv C-C\equiv C-(CH_2)_8-COO-(CH_2)_2$
$H_3C-(CH_2)_{12}-C\equiv C-C\equiv C-(CH_2)_8-COO-(CH_2)_2$ $\overset{+}{N}\overset{H}{\underset{(CH_2)_2-SO_3^-}{}}$ *22* 19, 23, 24)

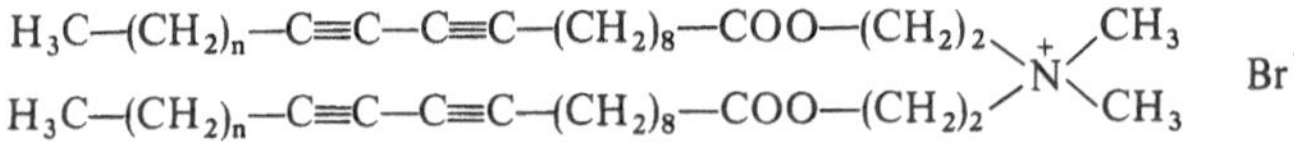

$H_3C-(CH_2)_{12}-C\equiv C-C\equiv C-(CH_2)_8-COO-(CH_2)_2-O-(CH_2)_2$
$H_3C-(CH_2)_{12}-C\equiv C-C\equiv C-(CH_2)_8-COO-(CH_2)_2-O-(CH_2)_2$ O *23*   19)

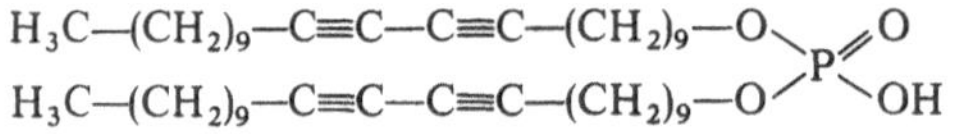

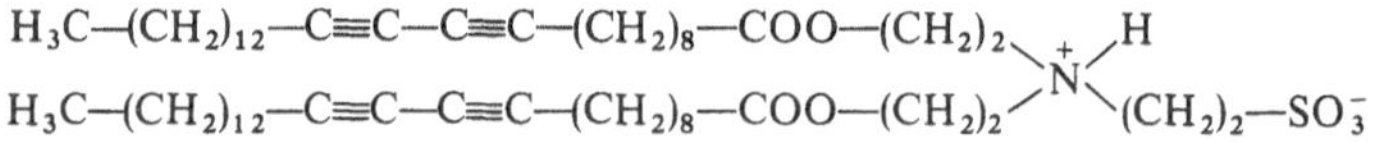

                          $H_3C-(CH_2)_{16}-COO-CH_2$
$H_3C-(CH_2)_{12}-C\equiv C-C\equiv C-(CH_2)_8-COO-CH$ *24*   19, 22)
                          $CH_2-O-PO_3H_2$

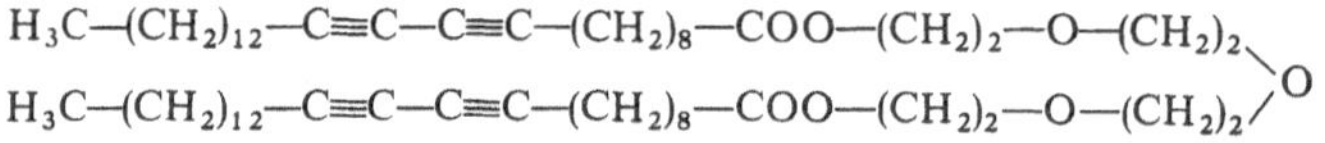

$H_3C-(CH_2)_{12}-C\equiv C-C\equiv C-(CH_2)_8-COO-CH_2$ *25*   22, 25)
$H_3C-(CH_2)_{12}-C\equiv C-C\equiv C-(CH_2)_8-COO-CH$

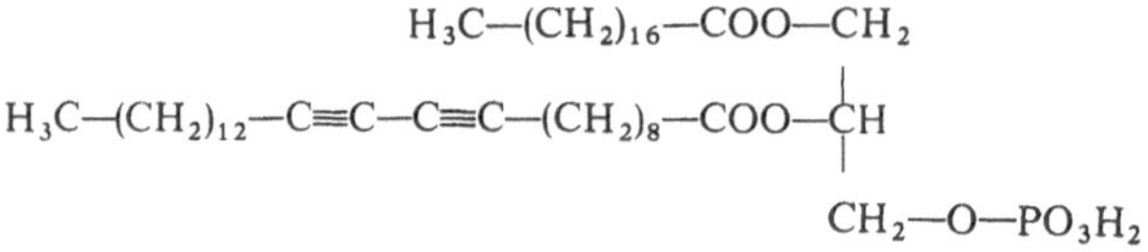

$$CH_2-O-\overset{O}{\underset{O_-}{\overset{\|}{P}}}-O-(CH_2)_2-\overset{+}{N}H_3$$

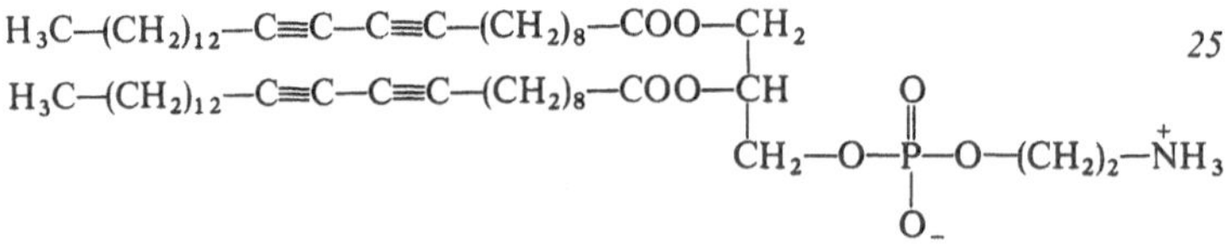

$H_3C-(CH_2)_{12}-CH=CH-CH=CH-COO-(CH_2)_2$
$H_3C-(CH_2)_{12}-CH=CH-CH=CH-COO-(CH_2)_2$ $\overset{+}{N}\overset{CH_3}{\underset{CH_3}{}}$ Br^- *26* 7, 25)

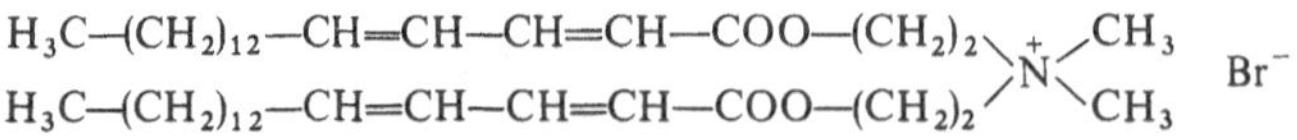

Table 1. (continued)

Type	Surfactant		Ref.

d)　$H_3C-(CH_2)_{17}-O-CH_2$
　　$H_3C-(CH_2)_{17}-O-CH$
　　　　　　　　　　$CH_2-O-CO-NH-(CH_2)_5-NH-CO-\overset{\overset{CH_3}{|}}{C}=CH_2$　　　27　　7)

　　$H_3C-(CH_2)_{17}-O-CH_2$
　　$H_3C-(CH_2)_{17}-O-CH$
　　　　　　　　　　$CH_2\ \ O\ \ CO-\overset{\overset{CH_3}{|}}{C}-CH_2$　　　28　　7)

　　$H_3C-(CH_2)_{17}\diagdown\overset{+}{N}\diagup CH_3$
　　$H_3C-(CH_2)_{17}\diagup\ \ \diagdown(CH_2)_3-NH-CO-\overset{\overset{CH_3}{|}}{C}=CH_2$　　Br^-　　29　　15, 26)

　　$H_3C-(CH_2)_{17}-O-CO-CH_2$
　　$H_3C-(CH_2)_{17}-O-CO-\overset{|}{CH}-NH-CO-(CH_2)_5-NH-CO-\overset{\overset{CH_3}{|}}{C}=CH_2$　　30　　7)

　　$H_3C-(CH_2)_{10}-COO-(CH_2)_2\diagdown\overset{\oplus}{N}\diagup CH_3$
　　$H_3C-(CH_2)_{10}-COO-(CH_2)_2\diagup\ \ \diagdown CH_2-CH=CH_2$　　$Br^\ominus$　　31　　12)

　　$H_3C-(CH_2)_{17}-O-CO-CH_2$
　　$H_3C-(CH_2)_{17}-O-CO-\overset{|}{CH}-NH-CO-\overset{\overset{CH_3}{|}}{C}=CH_2$　　32　　7)

　　$H_3C-(CH_2)_{14}-CO-O-CH_2$
　　$H_3C-(CH_2)_{14}-CO-O-\overset{|}{CH}$
　　　　　　　　　　$CH_2-O-CO-\overset{\overset{CH_3}{|}}{C}=CH_2$　　33　　7)

　　$H_3C-(CH_2)_{14}-COO-(CH_2)_2\diagdown$
　　$H_3C-(CH_2)_{14}-COO-(CH_2)_2\diagup N-CO-(CH_2)_2-\overset{\oplus}{N}\langle\text{pyridinium}\rangle\overset{\oplus}{N}-CH_2-CH=CH_2$　　$2\,Br^\ominus$　　34　　16)

　　$H_3C-(CH_2)_{17}\diagdown$
　　$H_3C-(CH_2)_{17}\diagup N-CO-(CH_2)_2-\overset{\oplus}{N}\langle\text{pyridinium}\rangle\overset{\oplus}{N}-CH_2-CH=CH_2$　　$2\,Br^\ominus$　　35　　16)

Table 1. (continued)

Type	Surfactant	Ref.

36

d) $H_3C-(CH_2)_{14}-COO-(CH_2)_2$
   $H_3C-(CH_2)_{14}-COO-(CH_2)_2$ N—CO—CH=CH—COO—(CH$_2$)$_2$—N⊕ ⟨ ⟩ ⟨ ⟩ N⊕—CH$_3$ Br⊖, I⊖   16,27)

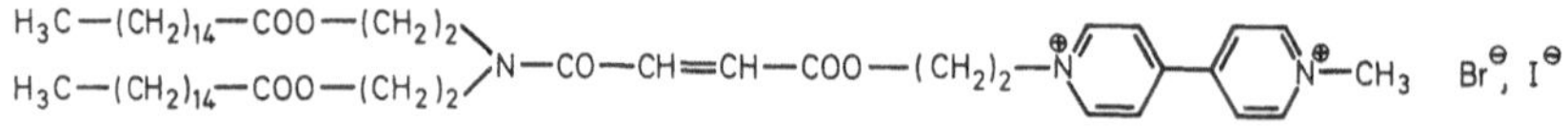

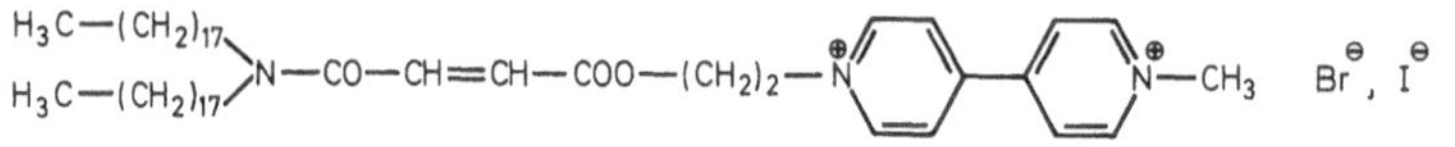

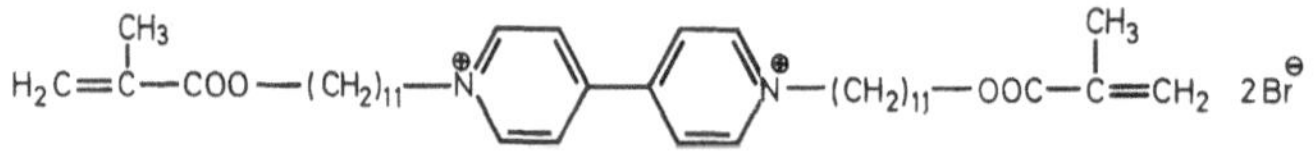

$H_3C-(CH_2)_{17}$
$H_3C-(CH_2)_{17}$ N—CO—CH=CH—COO—(CH$_2$)$_2$—N⊕ ⟨ ⟩ ⟨ ⟩ N⊕—CH$_3$ Br⊖, I⊖   16,27)

37

        CH$_3$ CH$_3$ 16)
$H_2C=\overset{|}{C}-COO-(CH_2)_{11}$—N⊕ ⟨ ⟩ ⟨ ⟩ N⊕—(CH$_2$)$_{11}$—OOC—$\overset{|}{C}$=CH$_2$ 2 Br⊖

38

In addition, the polycondensation of a 1:1 molar mixture of an amphiphilic diester (diethyl 2,2-dioctadecylmalonate (44) with an amphiphilic diamine (N,N'-dioctadecylethylene diamine (43)) was investigated in monolayers. [52]

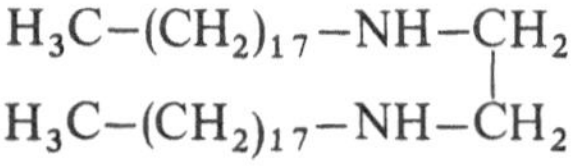

$$H_3C-(CH_2)_{17}-NH-CH_2 \qquad H_3C-(CH_2)_{17}\diagdown \qquad \diagup CO-O-C_2H_5$$
$$H_3C-(CH_2)_{17}-NH-CH_2 \qquad H_3C-(CH_2)_{17}\diagup C \diagdown CO-O-C_2H_5$$
$$(43) \qquad\qquad\qquad\qquad (44)$$

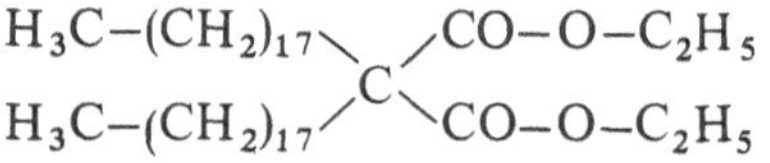

Details of these studies will be discussed in Section 3.2.3.

3 Polyreactions in Model Membranes

3.1 Membrane Models

Information about the bulk properties of lipids and surfactants is not sufficient to gauge their ability for forming membranes. The interaction with an aqueous phase results in totally different properties which can be described in terms of amphiphilic behavior, micelle and liposome formation, and formation of lyotropic phases [29]. For thermodynamic reasons lipids undergo self organization to form membrane like structures when brought in contact with an aqueous medium. Several methods have been developed in recent years for investigating these model membrane systems. The most common ones are monolayers at the gas/water interface [30], black lipid membranes [31], and vesicles (liposomes) [32] (Fig. 5 [33]).

In all three cases amphiphiles orient spontaneously to form structures resembling the phospholipid arrangement in biomembranes. These membrane models allow a variety of investigations of physical membrane properties which could not be conducted with the complex natural systems.

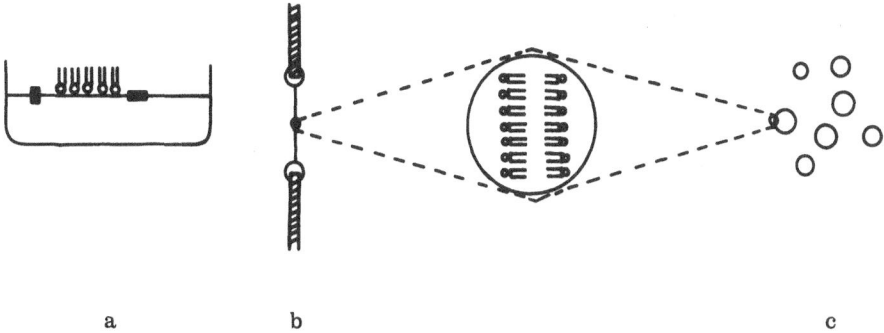

Fig. 5a—c. Orientation of amphiphilic compounds in model membranes [33]: **a)** monolayer at the gas/water interface; **b)** bimolecular lipid membrane (BLM); **c)** liposomes. Between b and c a cross section through the bilayer of BLM or liposomes is shown

3.2 Polymeric Monolayers

3.2.1 Investigation of Polymerizable Amphiphiles in Monolayers

The earliest reports on the behavior of oil on water were given by the Babylonians. They spread oil droplets on a water surface and used the behavior of the films for soothsaying on health, war, or wealth [34]. As early as 1917 Irving Langmuir recognized that amphiphilic substances can form a film on a water surface with a thickness of exactly one molecular layer [35]. Such a monomolecular film forms a two-dimensional system which, by variation of surface pressure, area and temperature, permits the measurement of a phase diagram. The most common form of recording these diagrams is the pressure/area isotherm measured with a Langmuir film balance (Fig. 6 [30]).

Various states can be seen in the monolayer in analogy to pV-diagrams of three-dimensional systems: at low pressures a gas analogous phase (a) can be formed which obeys a two dimensional gas law. Compression leads to an expanded or liquid

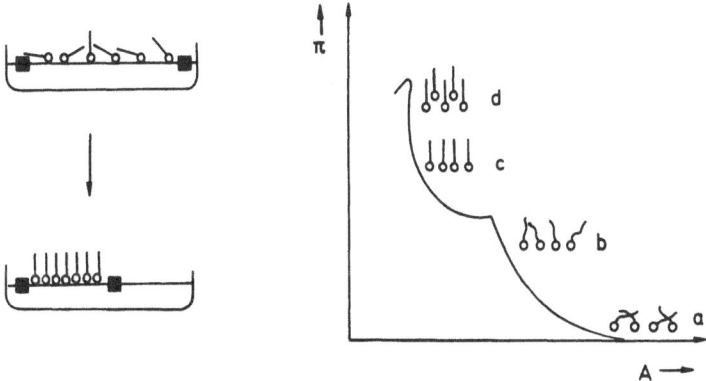

Fig. 6. Left: Langmuir film balance, schematic; right: surface pressure/area isotherm of a monolayer at the gas/water interface (**a–d**, see text). π: surface pressure; A: surface area [33]

analogous state (b) with contact of head groups, but high mobility of the hydrophobic chains. Further compression results in a condensed or solid-like phase with head packing (c) or chain packing (d). The smallest area of a solid film is in the order of 0.18 nm² per alkyl chain and corresponds to the occupied area of a paraffin chain in the crystal. Further decrease of surface area results in the collapse of the monolayer. Beyond this point the surface area decreases with constant surface pressure and the monolayer collapses into bilayers and multilayers.

The measurement of surface pressure/area isotherms provides a method for studying the influence of parameters such as temperature, head group charge and size, alkyl chain length, and pH on membrane properties. The effect of head group bulkiness on surface pressure/area isotherms of polymerizable lysophospholipid analogs is illustrated in Fig. 7 [36].

$$H_3C-(CH_2)_{12}-C\equiv C-C\equiv C-(CH_2)_9-O-\overset{\displaystyle |}{\underset{\displaystyle OH}{PO}}-OR$$

$$45, R = (CH_2)_2-\overset{+}{N}(CH_3)_3$$

$$46, R = (CH_2)_2-\overset{+}{N}H_3$$

$$47, R = H$$

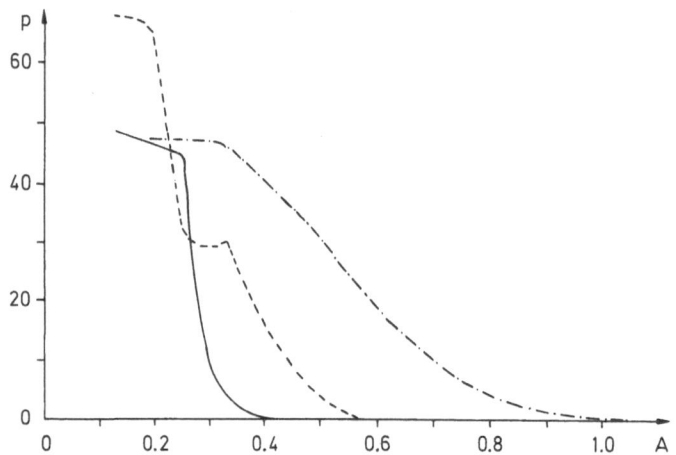

Fig. 7. Surface pressure/area isotherms of lysophospholipid analogous diacetylenic surfactants (45)–(47). (45), (—·—·—·—·—·); (46); (— — — — — —); (47), (————————). p: surface pressure in mN/m; A: area in nm²/molecule [33] (at 20 °C)

A decrease in occupied area of the head group results in an increase in packing density of the molecules: (45) exhibits only an expanded phase, (46) both a liquid and a solid-like phase, and (47) forms only a condensed film. Monolayer properties of many natural phospholipids and synthetic amphiphiles are described in the literature [37,38]. Especially the spreading behaviour of diacetylenic phospholipids at the gas-water interface was recently described by Hupfer [120].

Apart from the characterization of lipids in monolayers, several other examples for the versatility of this method have been described. Gorter and Grendel postulated the existence of a double layer in biomembranes from the occupied area of membrane extracts in monolayers [39]. Many membrane — mediated reactions can be monitored in monolayers, e.g. adsorption of proteins at a monolayer and insertion into the lipid matrix [40]. Another example is the investigation of the activity of the enzyme lipase as a function of the surface pressure [41]. Studying the miscibility of different phospholipids and the characterization of different phospholipid/cholesterol mixtures provides a method for the interpretation of their action in biomembranes [42] (see also 4.2.1). Interactions of local anesthetics with membranes have been studied in monomolecular films as a model [43].

The correlation of monolayer properties and spherical membranes (liposomes, cells) has been discussed by Blume [44].

3.2.2 Polymerization in Monomolecular Layers

Since the orientation and packing density of the monomers in monomolecular layers can be varied, this method allows the investigation of a polyreaction in such systems as schematically shown in Fig. 8 [45]. The orientation of monomer units remains unchanged during the polymerization process. The reaction results in a highly oriented, stable polymer film with a degree of orientation not achievable by other methods, e.g. by spreading a solution of the polymer at the gas/water interface. Irradiation-initiated monolayer polymerization has been studied intensively in recent years (for an extensive review of this subject cf. [46]). Normally the reactions proceed under contraction of the films. The final polymers occupy a smaller area, their compression isotherms exhibit steeper slopes and higher collapse pressure indicating lower compressibility and higher stability (Fig. 9 [47]).

In the case of absorbing monomers or polymers the polyreaction can be followed by UV/VIS spectroscopy. For this purpose, a special Langmuir film balance was constructed which fits in the sample chamber of a commercial UV spectrometer [48]. A polymerized monolayer of a diacetylenic lipid exhibits a strong absorption in the visible region. Thus, the time dependent increase in optical density is a measure for the course of the polyreaction. The absorption curves of such a polymerized monolayer [48] have the same characteristics as those of polymerized Langmuir-Blodgett multilayers of diacetylenic lipids [49]. As in the solid state [50], the polyreaction is topochemically controlled, i.e. the structure of the monomer crystal determines the reactivity and the structure of the polymer. The fully conjugated polymer is composed of alternating double and triple bonds (**Scheme 1.**). Its color (blue or red) depends on the conformation of the polymer backbone, i.e. on the effective conjugation number. The blue form (λ_{max}: 620 nm) is formed in the beginning of the polyreaction and corresponds to a large number of conjugated monomer units. The red form (λ_{max}: 540 nm) appears after prolonged irradiation or after heating the blue form above the phase transition temperature of the monomeric lipid. The packing density of the

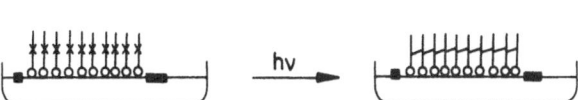

Fig. 8. Schematic of the preparation of polymeric monolayers by UV irradiation with retention of the orientation of the molecules (X = polymerizable group) [33]

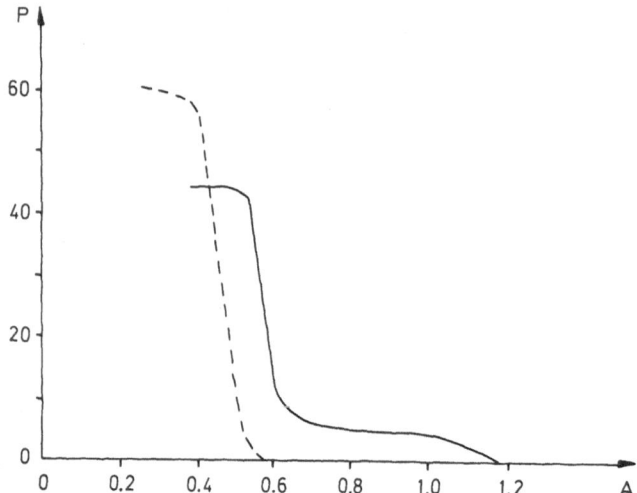

Fig. 9. Surface pressure/area isotherms of monomeric (————) and polymeric (— — — — —)
ammonium lipid (*20*, n = 12). p: surface pressure in mN/m; A: area in nm²/molecule [33]

R–C≡C–C≡C–$\overset{+}{\underset{R'}{N}}$–R–C≡C–C≡C–R' ⟶ $\overset{R}{\underset{R'}{C}}$–C≡C–C–$\overset{R}{\underset{R'}{C}}$–C≡C–C'   **Scheme 1.** Formation of poly-
(diacetylene)s.

monomers required for topochemical polymerization can easily be realized in
monomolecular layers. A polyreaction is only possible in a solid analogous monomer film, i.e. in the case of (20) (Fig. 9) at surface pressures above 10 mN/m [23,47].
Irradiation of a liquid film (p < 10 mN/m) does not lead to the formation of
polymer.

The fact that the polyreaction of diacetylenes is topochemically controlled is
especially well documented by the polymerization behavior of the sulfolipid (22) [23].
(22) forms two condensed phases when spread on an acidic subphase at elevated temperatures (Fig. 10). UV initiated polymerization can only be carried out at low
surface pressures in the 'first' condensed phase, where the molecules are less densely
packed. Apparently, in the 'second' phase at surface pressures from 20 to 50 mN/m
the packing of the diyne groups is either too tight to permit a topochemical
polymerization or a vertical shift of the molecules at the gas/water interface causes
a transition from head packing to chain packing (Fig. 10), thus preventing the formation of polymer.

In contrast to the topochemically polymerizable diacetylenes, methacryloylic and
butadienic lipids are also polymerizable in the liquid like phase. Figure 11 shows
the contraction behavior of butadiene lecithin (19) during irradiation in the liquid
analogous state [22]. The final polymer has a 1,4-trans structure [22].

Compared to diacetylenes, methacryloylic and butadienic lipids exhibit a higher
mobility of their polymer chains, and therefore, are more suitable for the formation

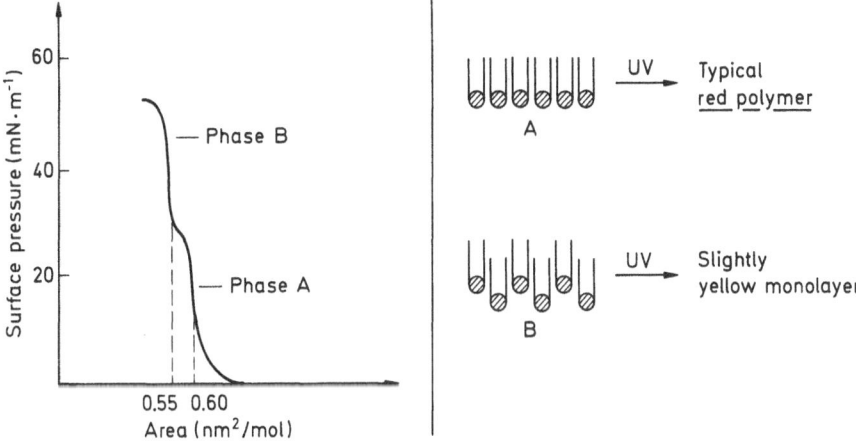

Fig. 10. Surface pressure/area isotherm of (*22*) at 41 °C, pH 2 [23]. Topochemical polymerization is impossible in the more densely packed phase (B) where most probably a change from head packing to chain packing of the lipid molecules has taken place

of flexible membranes. The surface pressure/area isotherms of all polymerized films exhibit a higher packing density and collapse pressure than those of the corresponding monomers, thus indicating a higher film stability of the polymerized membranes. The increased film stability is also demonstrated by a double layer of a poly-(diacetylenecarboxylic acid) which spans a width of 0.5 mm in diameter in an electron microscope grid. In contrast to unpolymerized monolayers which under these conditions would be destroyed within a few seconds [51], this bilayer membrane remains stable in water and air for weeks.

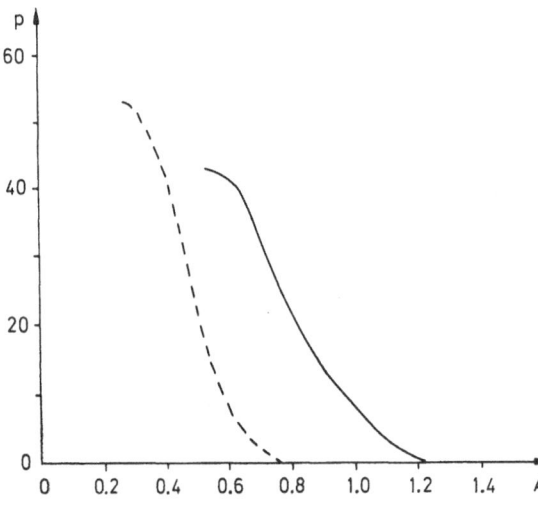

Fig. 11. Surface pressure/area isotherms of monomeric (————) and polymeric (— — — —) butadienic lecithin (*19*). p: surface pressure in mN/m; A: area in nm²/molecule [22]

3.2.3 Polycondensation in Monolayers

Besides the polymerization of diacetylenes, butadienes and acrylates, the polycondensation behavior of long-chain α-amino acids, diamines and diesters has been investigated in monolayers [52].

The amphiphiles (39)–(44) form either liquid or solid analogous monolayers,
depending on head group charge and chain length. Surface pressure/area isotherms
of α-amino acid esters (39)–(42) also vary strongly with the pH of the subphase.
The changes in compression behavior are brought about by ionization of the amino
groups on acidic subphases resulting in a less dense packing and a lower collapse
pressure than on alkaline subphases. Polycondensation is carried out under constant
surface pressure by monitoring the area change. After 24 h the films are allowed to
relax for 15 min and a surface pressure/area isotherm is recorded. In Figs. 12 and 13
the compression isotherms of monomers (40) and (42) are compared with the corresponding polymers [52]. In both cases polycondensation is accompanied by a film
expansion leading to less densely packed monolayers. Polypeptide formation is
illustrated in **Scheme 2**.

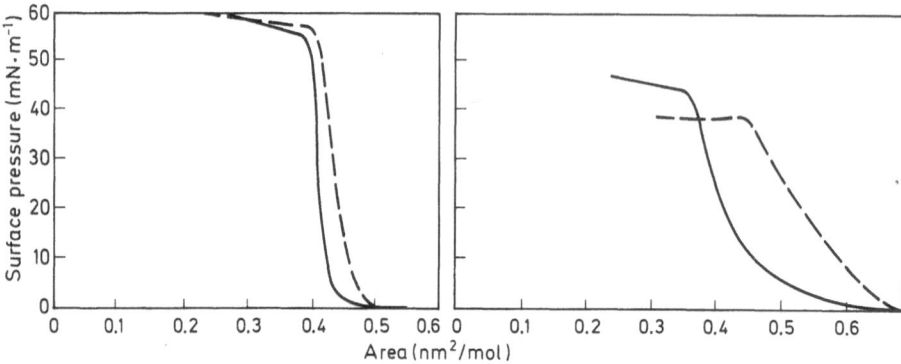

Fig. 12   Fig. 13

Fig. 12. Surface pressure/area isotherms of monomeric (————) and polymeric (40) (— — — —).
Polycondensation conditions: pH 8.5; 20 °C; 24 h; surface pressure: 2 mN/m [52]

Fig. 13. Surface pressure/area isotherms (30 °C) of monomeric (————) and polymeric (42)
(— — — — —). Polycondensation conditions: pH 8.5; 30 °C; 24 h; surface pressure 5 mN/m [52]

For the mixture of diamine (43) with diester (44) miscibility is necessary for a
successful polycondensation. Miscibility was demonstrated by showing that the mixed
monolayer obeyed Crisp's rule [53] (see 4.2.1). For a 1:1 molar mixture of diamine
(43) and diester (44) only a slight area change during monolayer polycondensation
could be observed. The compression isotherm of the polymer film exhibits a
diminished collapse area and pressure (Fig. 14). The structure of the final polymer is
shown in **Scheme 3**.

All polycondensation reactions were followed by FT-IR-spectroscopy. Polyamide
formation could be demonstrated by the disappearance of the ester $C=O$ stretching
mode and simultaneous formation of amide $C=O$ stretching modes [52]. Additional
proof for polymer formation is given by gel permeation chromatography. From the
elution volumes a molecular weight between 2,000 and 10,000 can be estimated.

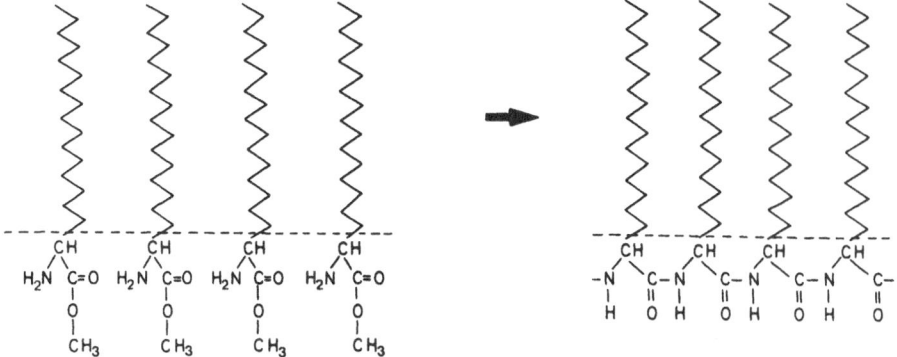

Scheme 2. Formation of an oriented polypeptide monolayer from monomeric (39) [52].

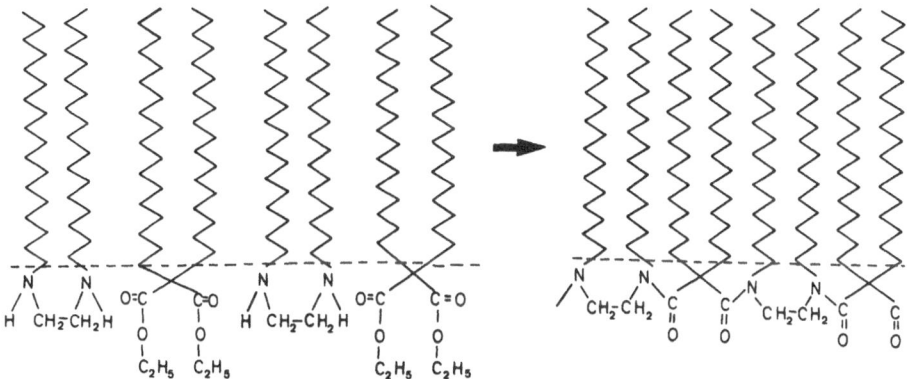

Scheme 3. Formation of an oriented polymeric monolayer via polycondensation of (43) and (44) [52].

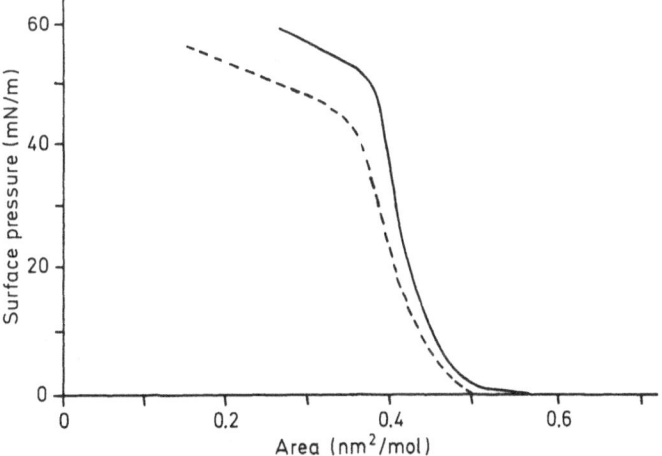

Fig. 14. Surface pressure/area isotherms of 1:1 molar mixture of (43) and (44). (————),
monomer, 20 °C, pH 8.5); (— — — — —), after 24 h at 30 °C, pH 8.5, surface pressure 5 mN/m [52]

The methyl esters (39) and (41) form polycondensates of higher molecular weights than the corresponding docosanyl esters (40) and (42).

The results illustrated so far show that polyreactions in oriented planar monolayers are possible and lead to highly ordered and very stable model membranes. This raises the question whether polymerization is also possible in bilayers (like vesicles and BLM) and whether these bilayers exhibit a higher stability than their unpolymerized counterparts.

3.3 Polymeric Bimolecular Lipid Membranes

Solvent-containing bimolecular lipid membranes are most frequently prepared by brushing a solution of appropriate lipids in e.g. decane over a hole (*ca.* 1 mm diameter) in a teflon foil. The support is located in a measuring cell containing water or electrolyte solution [31]. First, a rather thick lamella is formed. Interference colors can be observed in reflected light when the support is observed with a microscope (Fig. 15a [26]). Within seconds or minutes black dots appear in the colored membrane which eventually grow together (Figs. 15b and c), until the whole membrane appears black because of destructive interferences. In this state, the membrane consists of an oriented bimolecular film formed by self—organization of the lipid molecules. Since electrodes can be installed in the measuring cells separated by the BLM, these model systems are used for studying electrical properties of lipid bilayers and the effects of incorporation of ion carriers [31].

Unfortunately, BLMs are by far the least stable system among the membrane models illustrated in Fig. 5. In general, they can only exist when the lipid molecules are in the liquid-crystalline state. When a fluid BLM is cooled below the phase transition temperature of the lipid, the membrane is disrupted instantly. Moreover, even fluid

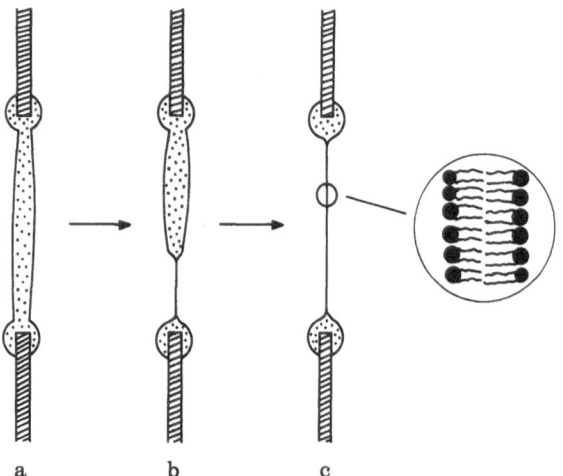

a   b   c

Fig. 15a–c. Formation of a bimolecular lipid membrane [26]. **a)** colored membrane (lamella), **b)** blackening membrane, **c)** black membrane

BLMs have lifetimes not exceeding minutes or hours. Therefore, it was desirable to build up black lipid membranes of high long-term stability.

Out of a variety of polymerizable lipids tested for possible use of bilayer formation, only three systems exhibited BLM lifetimes of more than a few minutes (Table 2 [26]). These BLMs were characterized by measuring their resistance and capacitance (Table 2., see [26] for details). The data obtained were comparable with values obtained with egg lecithin the most frequently used material for preparing BLMs.

Table 2. Lifetime, conductivity, and capacitance of BLM made from polymerizable amphiphiles [26]

Compound	$T/°C$	lifetime/h	$R/\Omega \cdot cm^2$	$C/\mu F \cdot cm^{-2}$
egg lecithin	20	2–3	10^7–10^8	0.33 ± 0.02
$(50)^a$	25	1–3	10^7–10^8	0.31 ± 0.02
(48)	40	1–2	10^7	—
(29)	21	5–10	10^6	0.38 ± 0.04

^{a} $H_3C-(CH_2)_{12}-C\equiv C-C\equiv C-(CH_2)_8-CO-O-(CH_2)_2-N\begin{smallmatrix} CH_2-CO \\ \\ CH_2-CH_2 \end{smallmatrix}\rangle O$ Ref. [36]

UV initiated polymerization conducted with fluid BLMs made from (48) and (50) failed. This can be explained by the fact that topochemical polymerization of diacetylenic lipids can only be carried out in the solid state (see 3.2.2). Attempts to cool diacetylenic BLMs below the phase transition temperature of the lipids used led to rupture of the membranes. In contrast to this, polymerization of the methacryloylic lipid (29) can be conducted in membranes which are in the fluid state. Since the unsensitized UV initiated polymerization of methacrylamides is very slow, methyl azoisobutyrate was added as initiator. The course of the polyreaction was followed by the change of charging current of the membrane during capacitance measurements (Fig. 16 [26]). During UV irradiation the membrane stayed black. Simultaneously, its capacitance rose from $0.38 \, \mu F \cdot cm^{-2}$ to $2 \, \mu F \cdot cm^{-2}$ corresponding to the slower current decrease of the oscilloscope trace (Fig. 16b). The current did not return to zero even after only 1 min of irradiation and increased continuously during the entire irradiation period (20 min, Fig. 16c). This corresponds to a decrease of membrane resistance from $10^6 \, \Omega \cdot cm^2$ to ca. $220 \, \Omega \cdot cm^2$. The latter value is only 20% higher than the resistance of pure electrolyte. This result could be explained by changes in the structure of the polymerized BLM (Fig. 16, left). During the polyreaction a contraction of the lipid head groups takes place. The membrane becomes "thinner" in some areas and water enters the hydrophobic part of the bilayer. This causes an increase of the dielectric constant leading to higher membrane capacitance. The contraction of the membrane eventually leads to the formation of holes (Fig. 16c). Since electrolyte can freely penetrate the bilayer, these holes are the reason for the low electric resistance after complete polymerization. Thus, the result of the reaction is a bilayer net which still exhibits a remarkably high mechanical stability. It must be emphasized again that the bilayer maintains its black color during UV irradiation pointing toward an intact BLM. The bilayer net

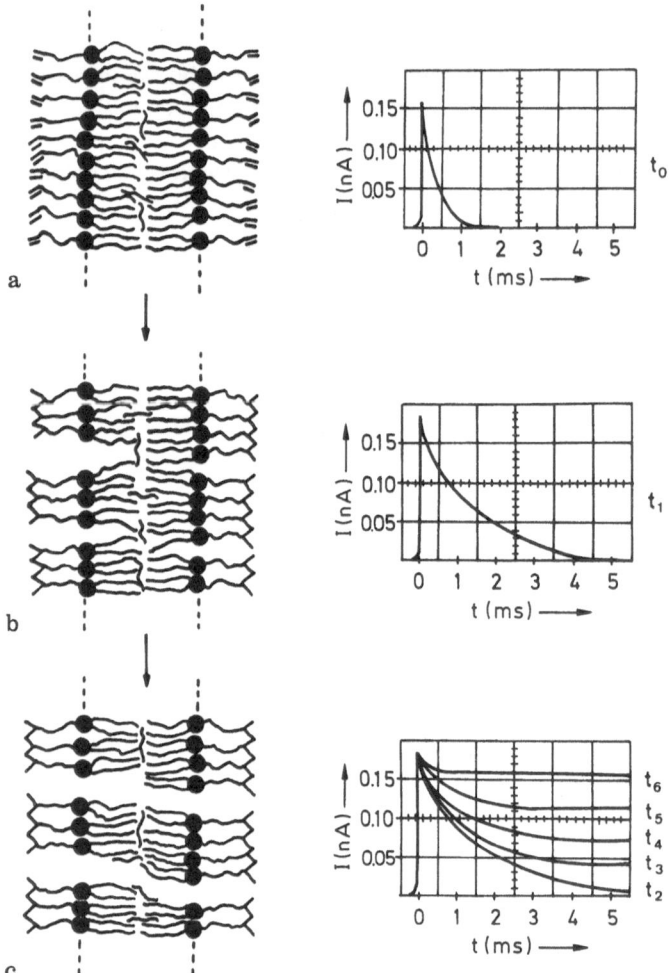

Fig. 16a—c. Polymerization of a BLM made from (29) [26]. Right: oscilloscope traces during capacitance measurements; left: molecular interpretation. a) monomeric membrane, b) membrane in the beginning of polyreaction, c) fully polymerized membrane. (t_2 = 1 min, t_6 = 20 min).

has a life time of up to 50 min. In contrast to the rapid rupture of an unpolymerized BLM, the breakdown of a polymerized black membrane proceeds at a very slow rate.

Since the desired long-term stability of polymerized BLM has not yet been accomplished, attempts are presently being made to prevent hole formation. This may be achieved by using mixtures of polymerizable and nonpolymerizable lipids for making BLMs, or by creating BLMs with higher solvent content. The viscosity of the membranes could possibly be diminished by these techniques so that the lateral diffusion of the lipid molecules becomes fast enough to prevent the appearance of holes. In addition, one could make use of a recently described method for preparing folded bilayers *below* the phase transition temperature of the lipid molecules [87]. In this case, the phase transition is accompanied by a contraction of the lipid head groups and an increase in membrane thickness. Possibly, polymerization of such a solid densely packed BLM will even increase membrane stability. Another way of stabilizing BLMs by using preformed hydrophobic polymer was accomplished by R. C. Hider et al. [88,89]

3.4 Polymeric Liposomes

3.4.1 Structure and Formation of Liposomes

Liposomes are the closest approach to biomembranes. They are closed spherical structures having an aqueous interior and one or several lipid double layers [54] (Fig. 17). Vesicles can be formed from synthetic amphiphiles and membrane extracts. Reconstituted membranes, i.e. liposomes from cell membrane constituents, contain nearly all components of cell membranes (lipids, proteins, glycolipids, etc.). Erythrocyte ghost cells formed by osmotic shock treatment represent another example for a model membrane whose composition corresponds to that of the membrane of living cells [55]. Vesicles made from synthetic lipids or lipid analogs have a much simpler composition and depending on the method of formation one obtains double or multilayered vesicles of different sizes [32].

The most common methods for preparing liposomes [32], which are not discussed here in detail, are the ultrasonication of lipid suspensions in water [6], the injection of alcoholic or etheral solutions of lipid into water, the dialysis of lipid-surfactant mixtures, and the removal of lipid films on glass surfaces by simple hand-shaking in water.

The polymerizable lipids and lipid analogs shown in Table 1 have been transformed into liposomal dispersions mainly by ultrasonication of their crystalline suspensions. Long-chain diacetylenecarbonic acids and their derivatives could be transformed into liposomes [23] in analogy to investigations of Gebicki and Hicks [57] with different unsaturated surfactants. Small and relatively homogeneous vesicles with a single bilayer are formed after prolonged sonication. Monomeric liposomes are relatively unstable. Like vesicles made from natural lipids, their solutions turn turbid after some days with precipitation of larger aggregates.

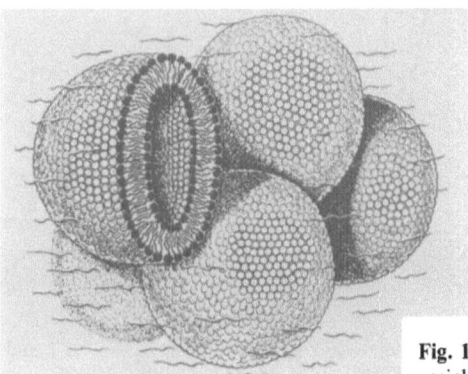

Fig. 17. Schematic representation of unilamellar lipid vesicles (liposomes) [33]

3.4.2 Polymerization in Liposomal Systems

Aqueous dispersions of polymerizable lipids and surfactants can be polymerized by UV irradiation (Fig. 18). In the case of diacetylenic lipids the transition from monomeric to polymeric bilayers can be observed visually and spectroscopically. This was first discussed by Hub [19] and Chapman [20]. As in monomolecular layers (3.2.2) short irradiation brings about the blue conformation of the poly(diacetylene) chain. In contrast, upon prolonged irradiation or upon heating blue vesicles above the phase transition temperature of the monomeric hydrated lipid the red form of the polymer is formed [23,120]. The visible spectra of the red form in monolayers and liposomes are qualitatively identical (Fig. 19).

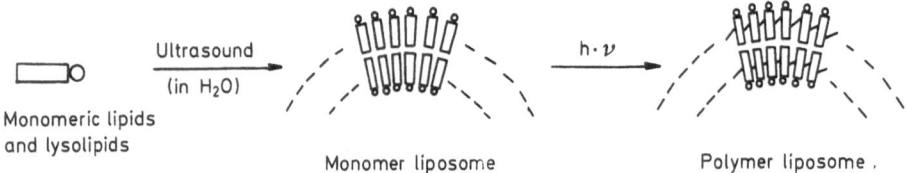

Monomeric lipids and lysolipids

Ultrasound (in H₂O)

Monomer liposome

h·ν

Polymer liposome.

Fig. 18. Scheme of formation of polymeric liposomes from polymerizable lipids [19, 33]

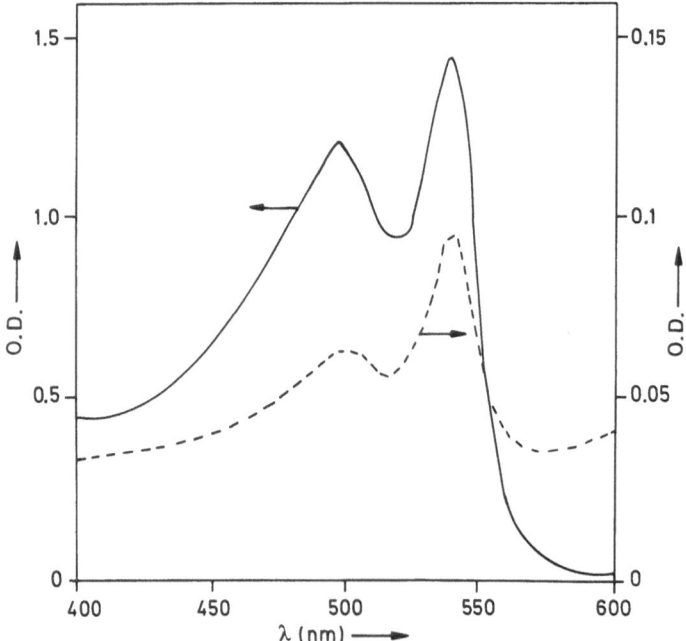

Fig. 19. Visible spectra of the red species of a polymerized monolayer (– – – – –) and polymerized liposomes (————) of diacetylenic lipid (*20*, n = 12) [19]. O.D.: optical density

In the case of butadienic lipids, the polyreaction was followed by a decrease of the strong monomer absorption (260 nm) [25]. Disappearance of vinyl protons in the ^{2}H NMR spectrum proved polymer formation for vesicles made from vinylic, acryloylic, and methacryloylic surfactants [13, 16].

Electron microscopy provides direct evidence that the polymerized dispersions still contain spherical liposomes (Fig. 20). Laser light scattering measurements also demonstrate the presence of vesicles in the irradiated dispersions [16]. The size distribution of vesicles is not significantly altered by polymerization [13] (Fig. 21) as confirmed by gel filtration before and after irradiation of vesicles of (7).

Polymerization of liposomes affects their stability. In contrast to monomeric liposomes, their polymerized counterparts remain stable for weeks. Entrapped substances are released to a much smaller extent from polymeric liposomes than from monomeric ones [13, 33]. This has been studied in the case of the butadienic lipid (19). Entrapped 6-carboxyfluorescein (6-CF) in high concentration exhibits self-quenching. Release into the surrounding aqueous medium results in a strong fluorescence due to dilution [59]. Below the phase transition temperature, liposomes made from dipalmitoyllecithin show an 8% release after 40 hours. Liposomes made from monomeric (19) are in the liquid state and release the dye more rapidly. Polymeric liposomes, however, show no significant release after 40 hours (Fig. 22 [33]). Entrapment and leakage of sugars from vesicles derived from (3), (5), (6), (7) and (19) were investigated too [13, 25].

An even more striking effect is observed by addition of the surfactant sodiumdodecyl sulfate (SDS) to vesicles (Fig. 23). While monomeric vesicles of (19) and dipalmitoyllecithin are destroyed by low SDS concentrations, the polymerized vesicles are stable up to $2 \cdot 10^{-3}$ mol/l SDS [25]. Polymerized vesicle dispersions can be diluted with ethanol without precipitation.[23] Polymeric liposomes of (20) are stable in 80% ethanol for weeks. This could also be shown by Regen et al. for polymerized vesicles of the methacryloylic lipids (4) and (6) [13, 14] (Fig. 24) by monitoring the turbidity (absor-

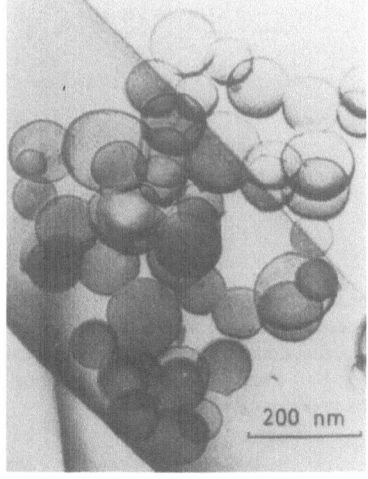

200 nm

Fig. 20. Electron micrograph of negatively stained (uranyl acetate) monomeric liposomes of diacetylenic lipid (20, n = 12) [19]

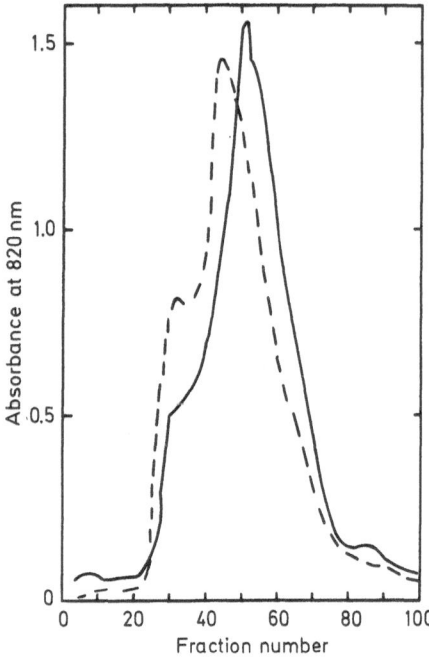

Fig. 21. Gel filtration of (7) (0.0006 mM) on cross-linked Sepharose 2B before (———) and after (— — —) polymerization. The mass balance for phosphorus (directly proportional to absorbance) was 92% and 91%, respectively. Elution was carried out with aqueous NaN$_3$ (0.02%), where 0.23 ml fractions were collected every minute. The void volume of the columns was contained in fractions 1–32. Reproducibility was ± 2 fractions [13]

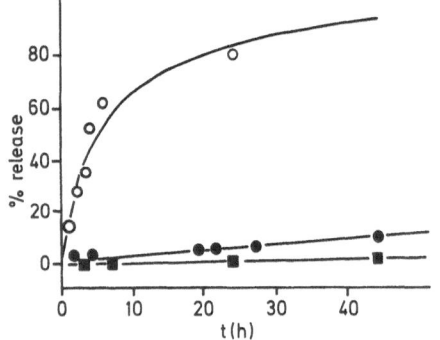

Fig. 22. Release of entrapped 6-carboxyfluorescein from liposomes of monomeric (○) and polymeric (■) (19) [33,25]. For comparison: dipalmitoylphosphatidylcholine vesicles (●) [25]

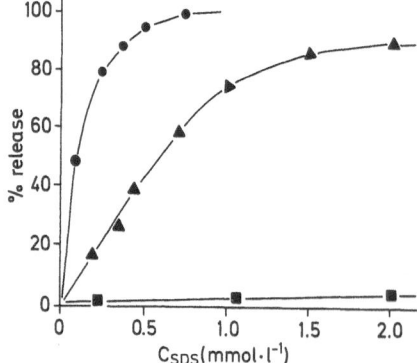

Fig. 23. Release of entrapped 6-carboxyfluorescein from liposomes of monomeric (●) and polymeric (■) (19) as a function of added sodium dodecyl sulfate (SDS). For comparison: dipalmitoylphosphatidylcholine vesicles (▲) [25]

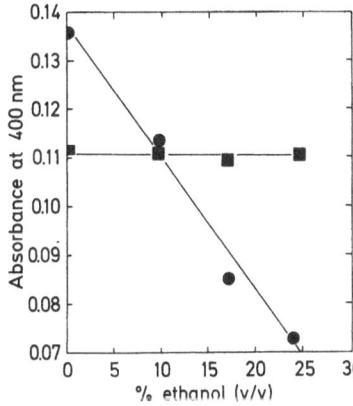

Fig. 24. Plot of absorbance at 400 nm as a function of percent ethanol (v/v) for polymerized (■) and nonpolymerized (●) vesicles of (4) [14]

bance at 400 nm) as a function of added ethanol. Within experimental error, the turbidity was constant from 0 to 25 % ethanol (v/v). In sharp contrast, similar experiments conducted with nonpolymerized vesicles showed a dramatic decrease in turbidity.

3.4.3 Phase Transitions in Polymeric Liposomes

So far, it has been shown that the stability of a model membrane can be tremendously increased by polymerization. This increased stability however, is associated with the presence of a polymer chain in the membrane itself or on its surface, bringing about increased viscosity and thus reduced flexibility. How does the reduced membrane mobility affect one of the most vital properties of biomembranes, the phase transition?

Whether polymerized model membrane systems are too rigid for showing a phase transition strongly depends on the type of polymerizable lipid used for the preparation of the membrane. Especially in the case of diacetylenic lipids a loss of phase transition can be expected due to the formation of the rigid fully conjugated polymer backbone [20] (Scheme 1). This assumption is confirmed by DSC measurements with the diacetylenic sulfolipid (22). Figure 25 illustrates the phase transition behavior of (22) as a function of the polymerization time. The pure monomeric liposomes show a transition temperature of 53 °C, where they turn from the gel state into the liquid-crystalline state [24]. During polymerization a decrease in phase transition enthalpy indicates a restricted mobility of the polymerized hydrocarbon core. Moreover, the phase transition eventually disappears after complete polymerization of the monomer [24].

In contrast, the phase transition of polymeric liposomes is retained if the polymer chain is more flexible or located on the surface of the vesicles instead within the hydrophobic core. Polymerized vesicles of methacrylamide (29) show a phase transition temperature which is slightly lower than the one for the corresponding monomeric vesicles (Fig. 26). This can be explained by a disordering influence of the polymer chain on the head group packing [15].

Deuteron -NMR measurements on the same system (labelled in the methyl group) also show that the mobility of the hydrophilic ammonium head group is reduced by

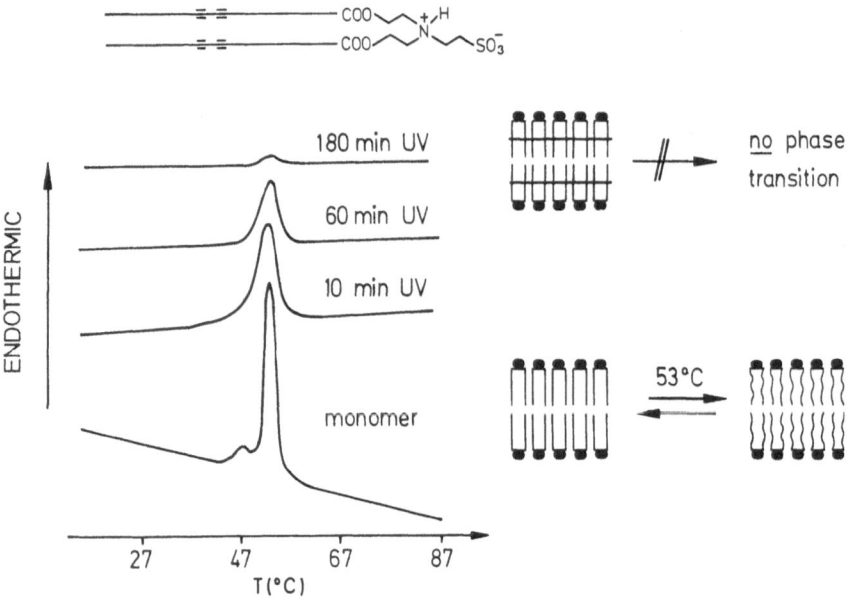

Fig. 25. DSC traces of vesicles of (22) after different polymerization times [24]

the polymer chain so that the "melting process" during the phase transition is restricted to the hydrophobic alkyl chains [59].

Methacryloylic lipid (5) is polymerizable in the hydrophobic part of the molecule. The phase transition temperature of the polymeric vesicle is again lowered compared to the non-polymerized vesicle (Fig. 27). The difference between the phase transition temperatures of monomer and polymer is somewhat larger than in the case of acrylamide (29). This might indicate that a saturated polymer chain in the hydrophobic core of a membrane decreases membrane order to a higher extent than a polymer chain on the membrane surface [15].

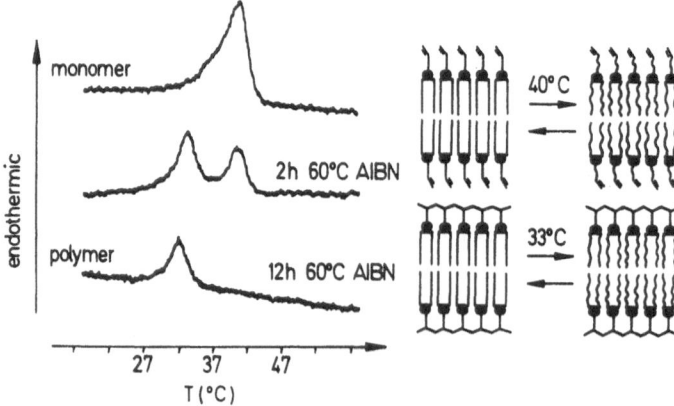

Fig. 26. DSC traces of vesicles of (29) after different polymerizations times

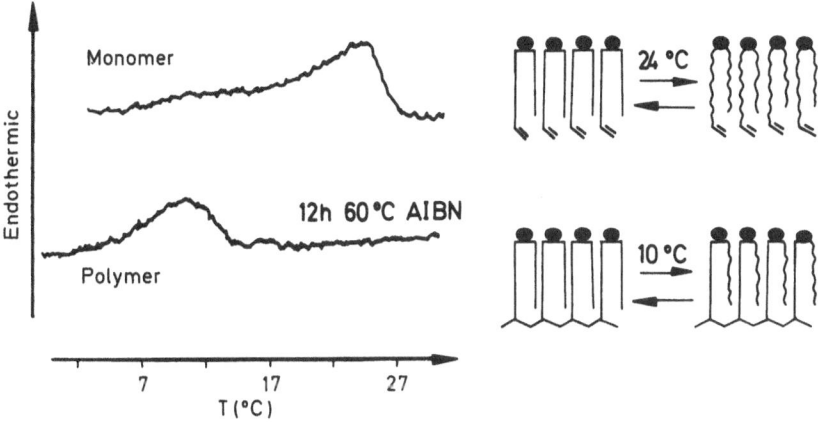

Fig. 27. DSC traces of vesicles of (5) before and after polymerization

3.4.4 Polypeptide Liposomes

The long-chain α-amino acid esters (40), (41), and (42) form bilayers on sonication in water under acidic conditions. Liposomes prepared from (40) and (42) precipitate if the aqueous medium is neutralized by titration with NaOH. Only liposomes made from (41) are stable even in basic solutions, as shown by electron microscopy [52]. Polypeptide formation in oriented spherical vesicles was confirmed by FT-IR spectroscopy. The liposomal solution of (41) was freeze-dried and the spectrum obtained from the residue was comparable with one of the polycondensed monolayers. The formation of polypeptide vesicles is illustrated in **Scheme 4**.

Polycondensation reactions in oriented monolayers and bilayers proceed without catalysis, and simply occur due to the high packing density of the reactive groups and their orientation in these layers. Bulk condensation of the α-amino acid esters at higher temperatures does not lead to polypeptides but to 2,5-diketopiperazines. No diketopiperazines are found in polycondensed monolayers or liposomes. Polycondensation in monolayers and liposomes leading to oriented polyamides represents a new route for stabilizing model membranes under mild conditions. In addition, polypeptide vesicles may be cleavable by enzymes in the blood vessels. In this case, they would represent the first example of stable but biodegradable polymeric liposomes.

3.4.5 Application of Polymerized Vesicles

Apart from the possible use of polymerized vesicles as stable models for biomembranes (Sect. 4) there may be a variety of different applications. Polymerized surfactant vesicles have been proposed to act as antitumor agents on a cellular level [33] in analogy to the action of the immune system of mammals against tumor cells [85]. Polymerized vesicles open the door to chemical membrane dissymmetry [22] which in turn, may lead to enhanced utility in photochemical energy transfer[84] (solar energy conversion, artificial photosynthesis). The utilization of unpolymerized lipo-

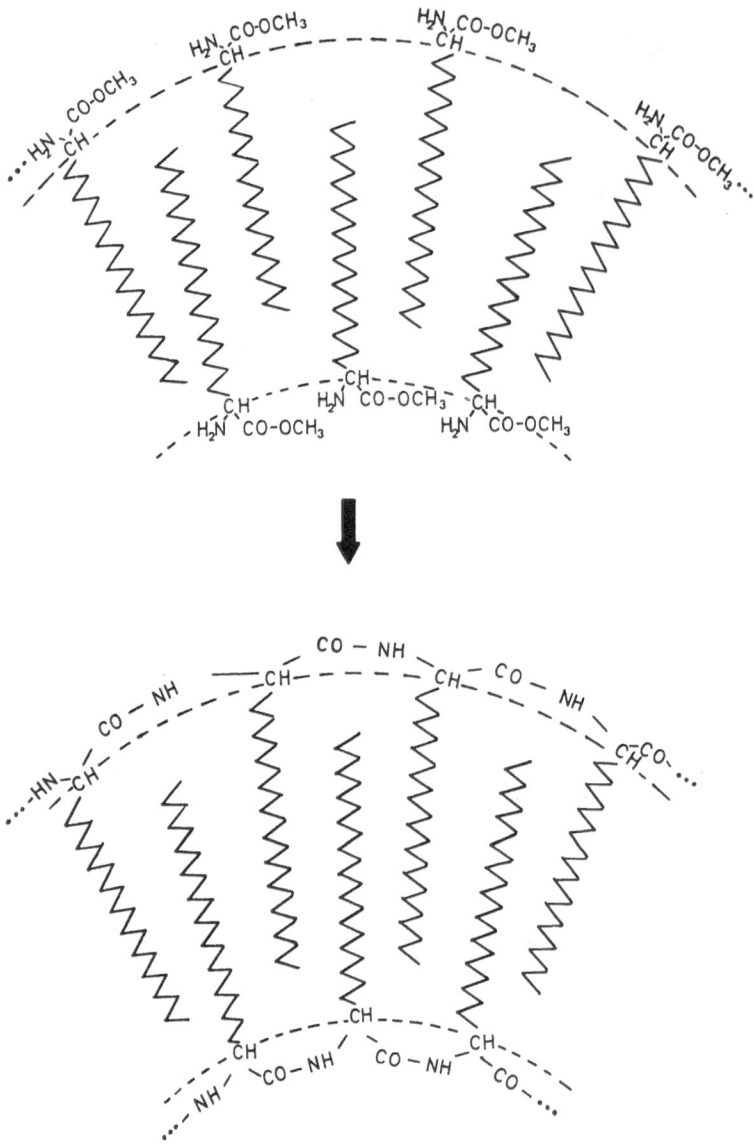

Scheme 4. Formation of polycondensed vesicles from monomeric (*41*) [52]

somes as target-directed drug carriers [34] through the blood circulation is hampered by their nonselective distribution, and their low stability when in contact with enzymes of the blood plasma. Polymerized vesicles could resist undesired enzymatic attacks and their distribution could be controlled by targeting them with a specific 'homing device', e.g. monoclonal antibodies. If this approach is feasible the limited biodegradability of most polymers utilized up to now for membrane stabilization might create another problem. Biodegradable polypeptide liposomes (3.3.4.) might be of better use for this particular purpose.

4 On the Way to Polymeric Biomembrane Models?

4.1 How to Approach the Problem

In general, model discussions have proven to be very helpful in modern cell membrane research [60]. Current membrane models — though perhaps oversimplified — are of high potential value for an understanding of important biological processes, e.g. intercellular communication, cell uptake of extracellular material, and transformation of external impulses into intracellular effects.

Thus far, it could be shown that stable liposomes can be prepared by polymerization of lipids. These vesicle systems, however, are still far away from being a real biomembrane model. As of now, they do not show any typical biological behavior such as surface recognition, enzymatic activities, variable lipid distribution, and the ability to undergo fusion.

There are two ways different in principle, to approach the problem of creating a polymeric biomembrane model. One can start out from a completely synthetic system and increase the similarity to natural systems by introducing natural lipids and

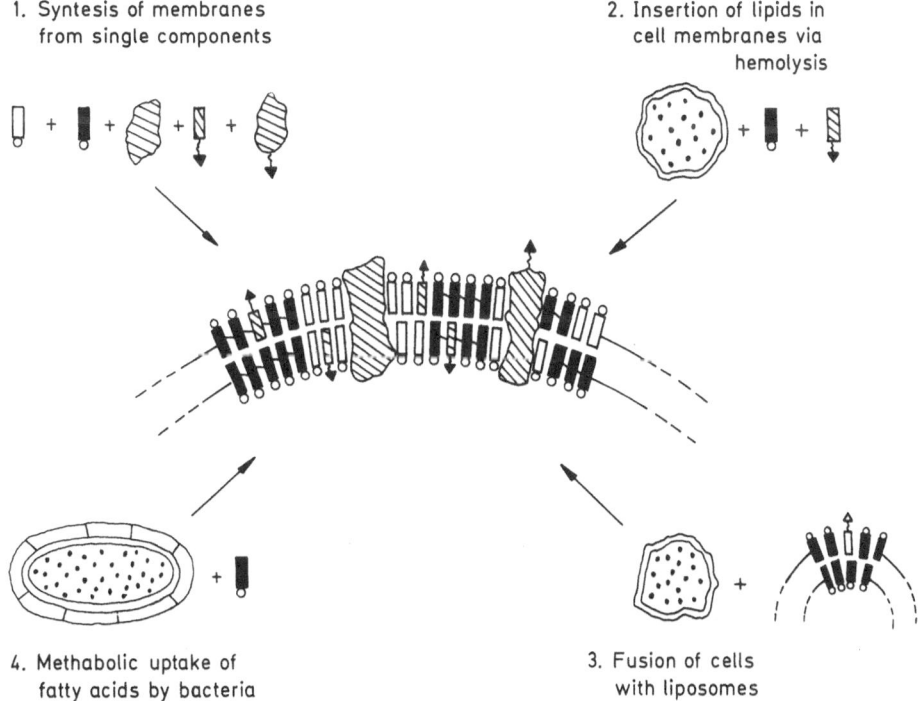

1. Syntesis of membranes
   from single components

2. Insertion of lipids in
   cell membranes via
   hemolysis

4. Methabolic uptake of
   fatty acids by bacteria

3. Fusion of cells
   with liposomes

Fig. 28. Schematic presentation of the build up of stable biomembrane models via partial polymerization of the membrane. Key: ⊏▭ , natural or synthetic lipids; ■▬○ , polymerizable lipids;  ,

proteins; ▧ ⊛ , lipids or proteins bearing cell recognizing groups [33]

membrane proteins. On the other hand, one can take a completely natural system which could be a living cell itself and introduce synthetic components into the cell membranes, e.g. polymerizable lipids. Figure 28 illustrates four possible paths to biomembrane models with polymerizable units.

First, a mixture of synthetic or natural phospholipids, polymerizable lipids, and proteins can be converted to liposomes and then be polymerized. Second, polymerizable lipids are introduced into e.g. erythrocyte ghost cells by controlled hemolysis and subsequent polymerization as described by Zimmermann et al. [61]. This hemolysis technique is based on a reversible dielectric breakdown of the cell membrane. Dielectric breakdown provides a third possible path to the production of biomembrane models. Zimmermann et al. could show that under certain conditions cells can be fused with other cells or liposomes [61]. Thus, lipids from artificial liposomes could be incorporated into a cell membrane. A fourth approach has been published by Chapman et al. [20]. Bacterial cells incorporate polymerizable diacetylene fatty acids into their membrane lipids. The diacetylene units can be photopolymerized in vivo. The investigations discussed in more detail below are based on approaches 1. and 3.

4.2 Synthetic Route to Polymeric Biomembrane Models

In this chapter emphasis will be placed on the preparation of polymeric membrane models using purely synthetic components or membrane constituents which are isolated in pure form from natural systems. Systems falling under this catagory are, for instance, mixed membranes of polymerizable and natural lipids or isolated membrane protein reincorporated into membranes of polymerizable lipids. Both the hydrolysis of the natural lipids by enzymes added to the vesicles preparation, or the remaining enzymatic activity of the reincorporated membrane protein in its new environment, can be used for testing the biological behavior of these functionalized membrane systems. Further biological functions can be introduced into polymerizable membranes by using lipids with recognizable hydrophilic head groups which provide sites for the surface recognition by proteins.

4.2.1 Incorporation of Natural Lipids into Polymerizable Membranes

What reasons are there for mixing polymerizable lipids with natural ones? Polymerized membrane systems, especially those based on diacetylenic lipids, have proven to be excessively rigid and to show no phase transition. Addition of natural lipids could help to retain a certain membrane mobility even in the polymerized state, with almost unaffected stability. Furthermore, natural lipids can provide a suitable environment for the incorporation of membrane proteins into polymerizable membranes (see 4.2.3). Besides this, enzymatic hydrolysis of the natural membrane component can be used for selectively opening up a vesicle in order to release entrapped substances in a defined manner (see 4.2.2). Therefore, it is interesting to learn about the miscibility of polymerizable and natural lipids and also about the polymerization behavior of these mixtures. Investigations on this subject have thus far focused on mixtures of natural lipids with polymerizable lipids carrying diacetylene moieties.

Their phase and polymerization behavior were characterized in monolayers and vesicles [62].

In general, the miscibility of two lipid components is influenced by different factors:
— size and charge of head groups [63]
— structure of alkyl chain [64]
— temperature [65]
— lateral membrane pressure [66]
— counter ions [67]
— phase transition temperature [64]

The cationic lipid (20, n = 12), the natural lipid (23) and the zwitterionic diacetylenic lecithin (18, n = 12) were mixed with naturally occurring lecithins, cephalins, and cholesterol.

Monolayers

Recording surface pressure/area isotherms is a relatively simple method for describing the miscibility behavior of a two component lipid mixture. The mean area per molecule of a binary mixture can be calculated using the following equation:

$$\bar{A} = x_1 A_1 + (1 - x_1) A_2$$

with $\bar{A}$ the mean area per molecule, x_1 the mole fraction of lipid 1, and A_1, A_2 the partial molar areas of the two lipids at the applied pressure. In the case of ideal miscibility or complete immiscibility A_1 and A_2 are constant and equal to the values of the pure components. Plotting $\bar{A}$ versus x_1 results in a straight line (Fig. 29, right). Any deviation from this linear relation indicates interaction between the molecules, thus non-ideal miscibility.

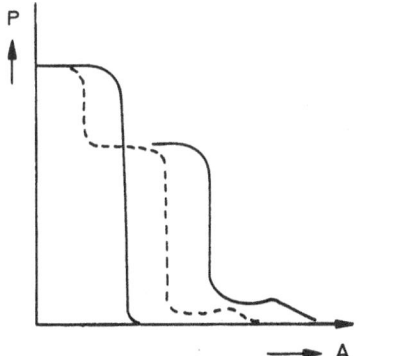 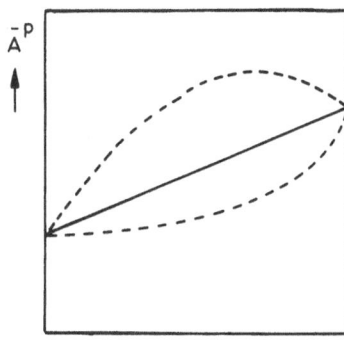

Fig. 29. Characterization of mixed monolayers; left: surface pressure/area isotherms of pure components (————) and of completely phase separated 1:1 mixed monolayer (— — — — —); right: plot of mean area per molecule at a certain surface pressure ($\bar{A}^P$) versus mole fraction (x) for completely non mixed or ideally mixed (————) monolayer. (— — — — —) illustrates positive or negative deviations from linearity indicating the presence of not ideally mixed films

Differentiation between ideal miscibility and complete immiscibility is possible by evaluating surface pressure/area isotherms. According to the phase rule of Defay and Crisp [53, 67] in a completely immiscible monolayer the surface pressures observed for phase transitions or collapse points are equal to those of the pure components. This case of a completely immiscible monolayer is schematically illustrated in Fig. 29 (left). In a completely miscible lipid monolayer these surface pressures vary with different molar ratios of the lipid components.

Due to the topochemical restrictions of diacetylene polymerization, diacetylenic lipids are solely polymerizable in the solid—analogous phase. During the poly-reaction an area contraction occurs leading to a denser packing of the alkyl chains. In addition to surface pressure/area isotherms the polymerization behavior of di-acetylenic lipids containing mixed films give information about the miscibility of the components forming the monolayer:

1) Provided the components are completely miscible and hexagonally packed in a mixed film below a molar ratio of 0.25 of diacetylenic lipid, each of the 6 nearest neighbors of a polymerizable lipid molecule is a nonpolymerizable natural lipid. Due to the low lateral diffusion rate in the condensed phase diacetylene poly-merization should either become impossible or at least proceed at a considerably lower rate.

2) In contrast, unchanged polymerization rates for all molar ratios of a polymeri-zable lipid indicate the formation of monomer islands and thus complete immisci-bility.

3) If miscibility results in the formation of a liquid — analogous phase, poly-merization of the diacetylenic component should no longer be possible.

Surface pressure/area isotherms of mixtures of the cationic lipid (20, n = 12) with distearoylphosphatidylcholine (DSPC) are shown in Fig. 30. For all mixtures only one collapse point is observed. The collapse pressure increases continuously with increasing amount of DSPC, indicating miscibility of the two components. Plotting A versus molar ratio (Fig. 31) results in considerable deviation from linearity, which also suggests miscibility of the two compounds in monolayers. This is also confirmed by the fact that the polymerization rate, as measured by the increase of optical density at 540 nm, is reduced by a factor of 100 when the DSPC molar ratio is increased from 0 to 0.52.

In contrast to this, the system neutral lipid (23)/DSPC shows considerably smaller deviations from the additivity rule and the surface pressure/area isotherms indicate two collapse points corresponding to those of the pure components [62]. Photopoly-merization can be carried out down to low monomer concentrations and no rate change is observed. These facts prove that the system (23)/DSPC is immiscible to a great extent. The same holds true for mixed films of diacetylenic lecithin (18, n = 12) with DSPC, as well as for dioleoylphosphatidylcholine (DOPC) as natural component.

Since cholesterol is an important component of many biological membranes mix-tures of polymerizable lipids with this sterol are of great interest. In mixed monolayers of natural lipids with cholesterol a pronounced "condensation effect", i.e. a reduction of the mean area per molecule of phospholipid is observed [68]. This influence of cholesterol on diacetylenic lecithin (18, n = 12), however, is not very significant (Fig. 32). Photopolymerization indicates phase separation in this system. Apparently due to the large hydrophobic interactions between the long hydrocarbon chains of

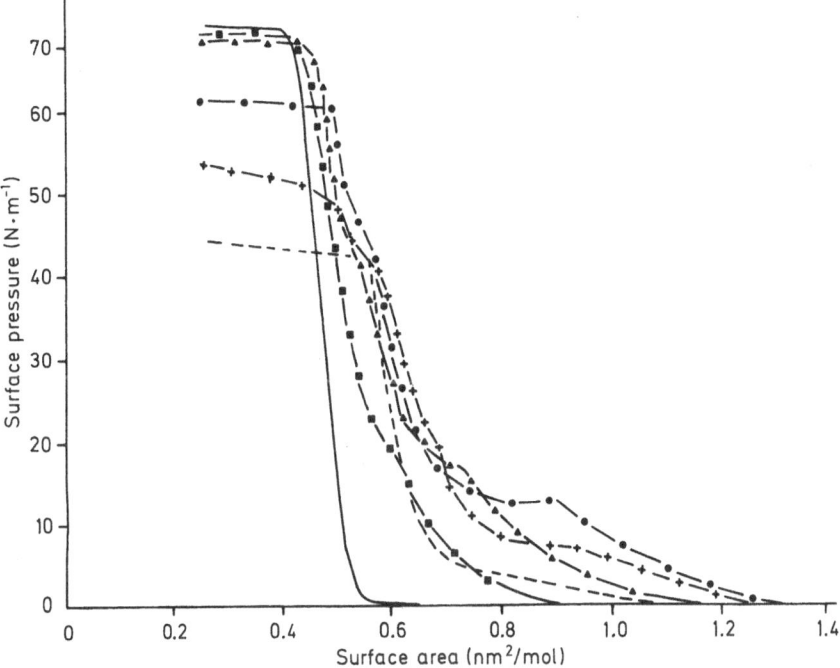

Fig. 30. Surface pressure/area isotherms of mixed monolyers of (*20*, n = 12) with distearoyl-phosphatidylcholine at 20 °C. Mole fraction of (*20*): 0, (————); 0.2, (■); 0.4, (▲); 0.6, (●); 0.8, (+); 1.0, (— — — — —)[62]

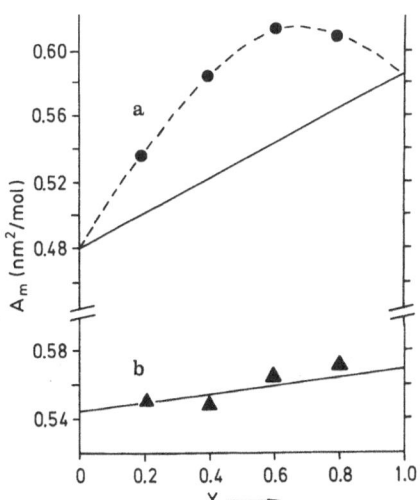

Fig. 31. Mean molecular area (A_m) as a function of the composition for mixed monolayers of di-stearoylphosphatidylcholine with (*20*, n = 12) (●, 30 mN/m) and (*23*) (▲, 15 mN/m) at 20 °C [62]

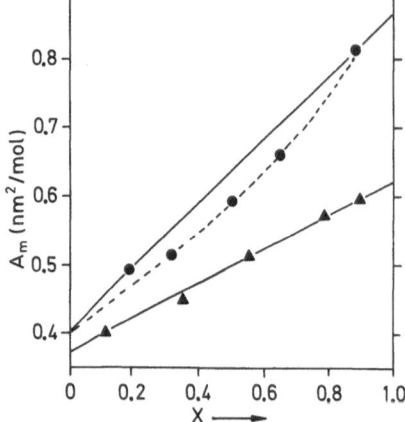

Fig. 32. Mean molecular area as a function of the composition for mixtures of (*18*, n = 12) with cholesterol at 12 mN/m (▲, 20 °C, ●, 40 °C) [62]

(*18*), this compound crystallizes separately. The same explanation was suggested for the immiscibility of the system dibehenoyllecithin/cholesterol [64]. Above the phase transition temperature the phospholipid (*18*) is miscible with cholesterol (Fig. 32).

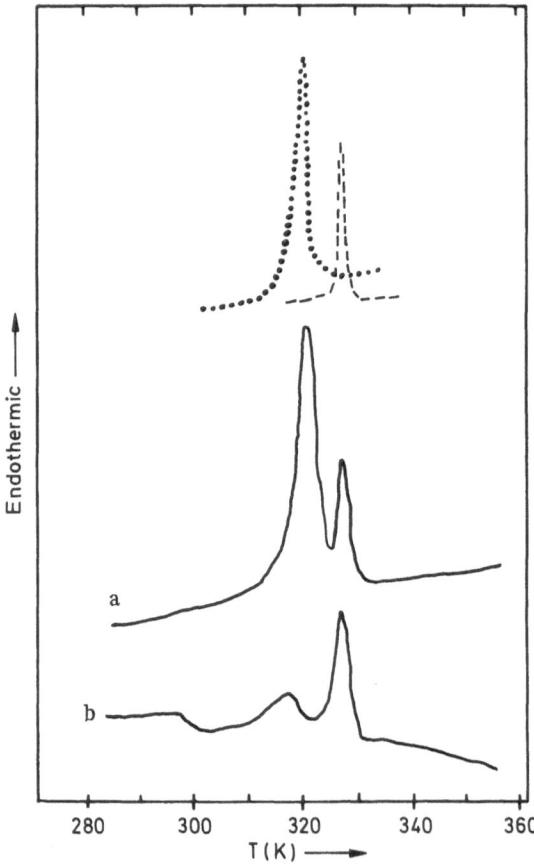

Fig. 33a and b. DSC traces of vesicles of (*23*), (. . . . .) and distearoylphosphatidylcholine (DSCP), (— — —); **a)** 1:1 molar mixture of (*23*) and DSPC, monomeric; **b)** after polymerization [62]

Vesicles

In mixed bilayer vesicles diacetylenic and natural lipids exhibit the same miscibility behavior as in monomolecular films. This can be demonstrated using differential scanning calorimetry (DSC). The neutral lipid (23) is immiscible with DSPC or DOPC as indicated by the two phase transitions of the mixed liposomes which occur at the same temperatures as those of the pure components (Fig. 33 a).

After vesicle polymerization the phase transition of the diacetylenic lipid has almost completely disappeared, while the phase transition of DSPC is unaffected by polymerization (Fig. 33 b). The same holds true for mixtures of (23) with DOPC [62].

DSC traces of mixtures of cationic lipid (20 with DOPC (Fig. 34) indicate strong interaction between the components. Upon heating, the unpolymerized vesicle dispersion exhibits four phase transitions none of which corresponds to the pure

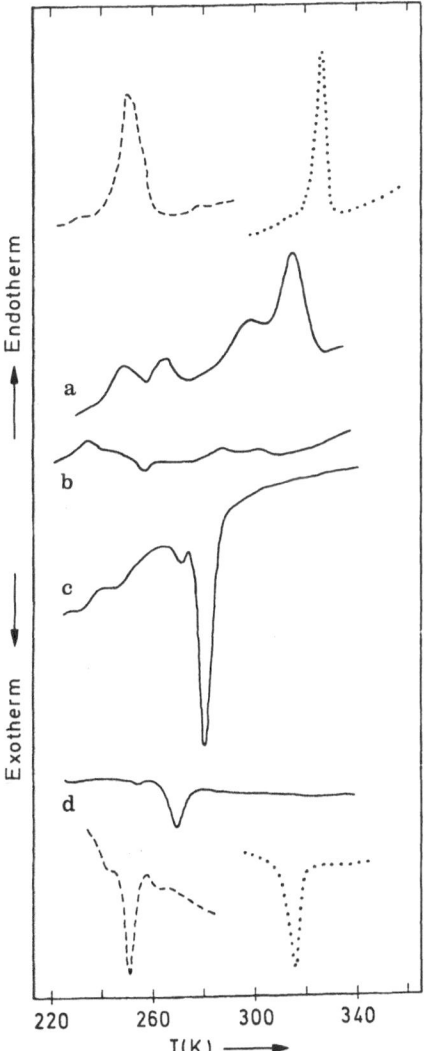

Fig. 34a–d. DSC traces of 1:1 mixed vesicles of (20) and dioleoylphosphatidylcholine (DOPC). (. . . . .), pure (20); (– – – – –), pure DOPC; a) heating run of monomeric mixture; b) heating run of polymeric mixture; c) cooling run of monomeric mixture; d) cooling run of polymeric mixture [62].

compounds, while upon cooling only one major transition can be observed. This may be explained by the existence of one homogeneous phase in the liquid crystalline state, whereas in the gel state there is a complex mixture of several phases of different composition. In contrast to the immiscible systems neutral lipid (23)/DSPC and lecithin (18, n = 12)/DOPC, polymerization of this mixture strongly reduces the mobility of the natural lipid and leads to decreased transition enthalpies.

Besides differential scanning calorimetry, electron microscopy can also serve for characterizing the mixing behavior of multicomponent vesicular systems. The "ripple structure" of phospholipids with saturated alkyl chains (also referred to as smectic Bca phase, Fig. 35) is taken to indicate patch formation (immiscibility) in mixed phos- close enough (1–2 nm) lipid molecules are able to diffuse from one membrane to the between the pre- and main-transition of the corresponding phospholipid, electron

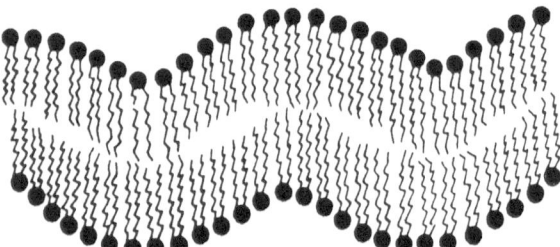

Fig. 35. Scheme of the "ripple structure of bilayers of saturated phospholipids found between the pre-transition and main-transition of the membrane [86]

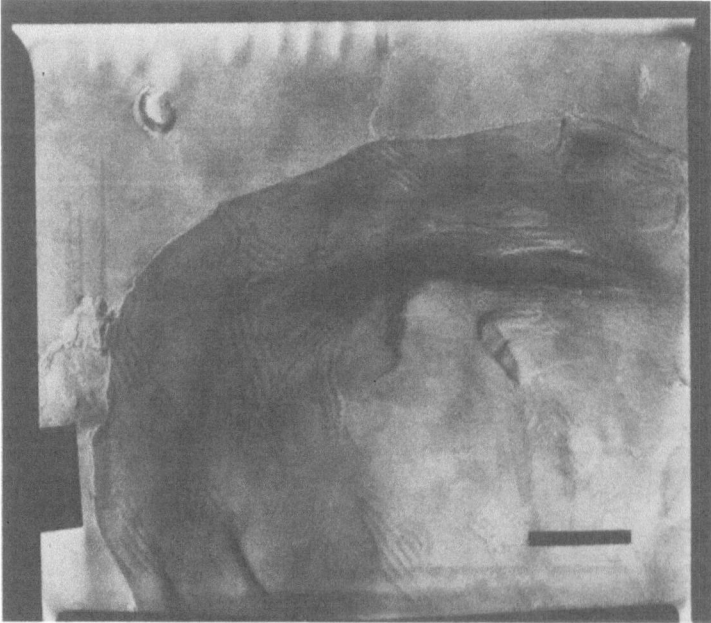

Fig. 36. Electron micrograph of ripple structure and patch formation in 1:1 mixed bilayers of (18, n = 12) and dimyristoylphosphatidylcholine. Bar represents 250 nm

microscopic sample preparation has to be carried out in a well defined temperature range. First experiments with the system diacetylenic lecithin (*18*, n = 12)/DMPC (dimyristoylphospatidylcholine) [69] indicate the existence of patches of ripple-structured phospholipid next to unstructured areas of diacetylenic lipid (Fig. 36). This points towards immiscibility of the two components. These miscibility experiments provide the basis for further investigations on the hydrolytic activity of enzymes on the natural membrane component, which can be used for selectively opening up liposomal systems.

4.2.2. Enzymatic Hydrolysis of Natural Lipids in Polymeric Membranes

There are several methods to selectively open up closed polymeric membrane compartments in order to release entrapped substances (Fig. 37). For 'uncorking' a polymerized vesicle, its membrane has to contain destabilizable areas which could possibly be opened up by variation of pH [70], temperature increase [71], photochemical destabilization [72], or enzymatic processes. Such an enzymatic process is the hydrolysis of a natural phospholipid by phospholipase A_2 (Fig. 38). This enzyme cleaves the ester bond in position two of a natural phosphoglyceride producing a lysophospholipid and a fatty acid which are both water soluble. This leads to complete destruction of the membrane.

The action of phospholipase A_2 on mixed monolayers of natural and polymerizable lipids can be measured under constant surface pressure by the contraction of the monolayer as a function of time as depicted schematically in Fig. 39. It turns out that the chief parameter influencing the enzymatic activity is the miscibility of the lipid components and not the fact whether the film is polymerized or not. In mixed and demixed membranes the enzyme is able to hydrolyze the natural lipid component, but with considerable differences in the hydrolizing rate (Fig. 40). A pure dilauroyllecithin (DLPC) monolayer is completely hydrolyzed in a few minutes after injecting the enzyme

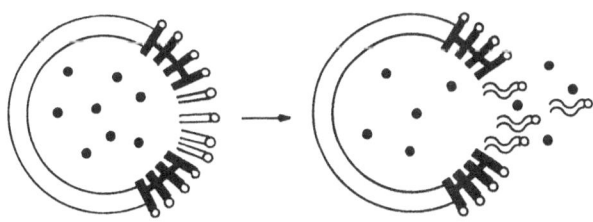

POSSIBLE "CORK-SCREWS":

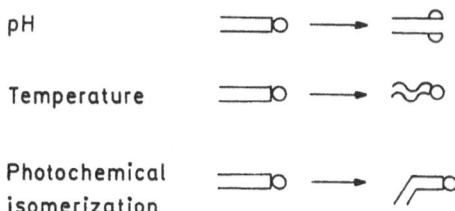

Fig. 37. Methods to release entrapped substances from partially polymerized mixed vesicles (schematic).

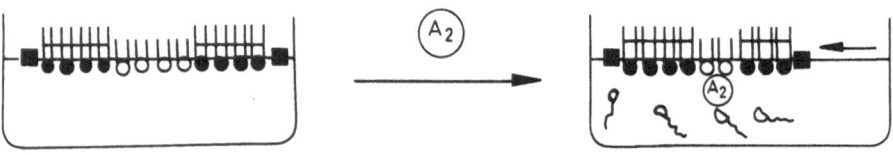

Fig. 38. Enzymatic hydrolysis of a phospholipid by phospholipase A_2

Immiscible system

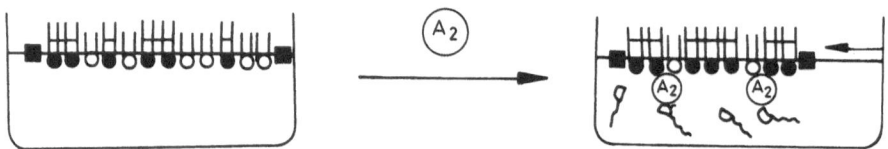

Miscible system

Fig. 39. Scheme of action of phospholipase A_2 on mixed monolayers of natural (○━) and polymerizable (●━) lipids [62]

into the subphase (a). For a 1:1 mixture of DLPC with an immiscible polymerizable lipid (b) the hydrolysis rate is slowed down and the final area is 50% of the original one as to be expected for a 1:1 mixture if one component cannot be hydrolyzed. In a miscible system (c) the hydrolysis rate is decreased to a very high extent. This can be explained by the fact that each DLPC molecule is encompassed by a number of nonhydrolyzable lipid molecules and therefore, the action of the enzyme might be hindered [62].

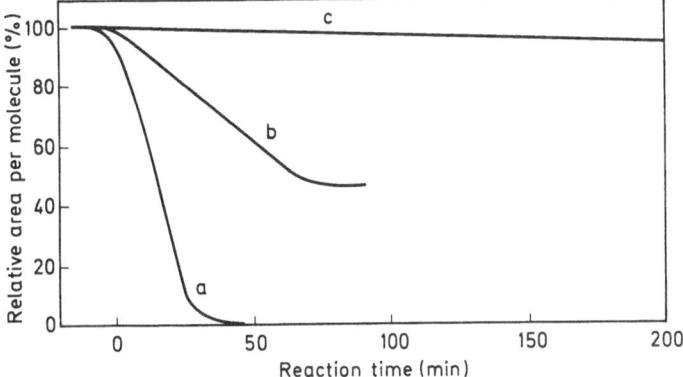

Fig. 40a—c. Relative area contraction of polymerized monolayers during enzymatic hydrolysis[62]. a) pure dilauroylphosphatidylcholine (DLPC); b) 1:1 mixture of (*23*) with DLPC (phase separated); c) 1:1 mixture of (*20*, n = 12) with DLPC (not phase separated)

Preliminary experiments with 6-CF loaded mixed vesicles of dipalmitoylphosphatidylcholine (DPPC) and diacetylenic lipid (*22*) exhibit only a very slow release of 6-CF after the addition of phopholipase A_2. Reasons for this may be headgroup interactions which decrease the enzymatic activity or the fact that multilamellar vesicles were used for this study. Further experiments will be carried out with "giant" liposomes (visible under the light microscope) which are unilamellar and provide a higher enclosure percentage and low surface curvature (see 4.3.2).

4.2.3 Incorporation of Membrane Proteins into Polymeric Membranes

Biological membranes are always pictured as being very selective barriers separating different biochemical reaction compartments. This high performance transport specificity solely depends on the presence of membrane proteins embedded in the lipid matrix. On the other hand, most membrane proteins cease to function in the absence of lipids. In order to introduce biological transport abilities into artificial membrane systems protein-lipid interactions are of vital interest. The question is how the activity of membrane proteins is affected if they are placed into a polymeric environment.

As an example of an asymmetric membrane integrated protein, the ATP synthetase complex (ATPase from Rhodospirillum Rubrum) was incorporated in liposomes of the polymerizable sulfolipid (22)[24]. The protein consists of a hydrophobic membrane integrated part (F_0) and a water soluble moiety (F_1) carrying the catalytic site of the enzyme. The isolated ATP synthetase complex is almost completely inactive. Activity is substantially increased in the presence of a variety of amphiphiles, such as natural phospholipids and detergents. The presence of a bilayer structure is not a necessary condition for enhanced activity. Using soybean lecithin or diacetylenic sulfolipid (22) the maximal enzymatic activity is obtained at 500 lipid molecules/enzyme molecule. With soybean lecithin, the ATPase activity is increased 8-fold compared to a 5-fold increase in the presence of (22). There is a remarkable difference in ATPase activity depending on the liposome preparation technique (Fig. 41). If ATPase is incorporated in-

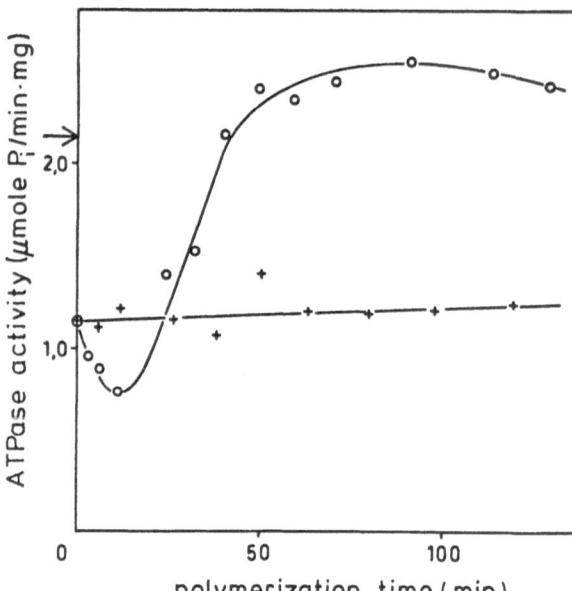

Fig. 41. ATPase activity of ATP synthetase incorporated into liposomes; (-○-) incubation of the enzyme with polymerized liposomes; (-+-) incubation of the enzyme with monomeric liposomes of (22) followed by polymerization; (→) activity of ATP synthetase in soybean lecithin liposomes [24]. ATPase activity was measured in 100 mM Tris-HCl (pH 8.0) in the presence of 1 mM ATP at 37 °C by determining the liberated orthophosphate

to already partially polymerized liposomes the hydrolytic activity exhibits a significant dependence on further polymerization time. A minimum of enzymatic activity after 10 min polymerization is followed by a strong increase (even above the activity in soybean lecithin liposomes) upon prolonged irradiation. The activity minimum corresponds with a red shift in the VIS spectra of the diacetylene polymer which is associated with a conformation change of the polymer. Incubation of the ATP synthetase complex with monomeric liposomes followed by irradiation only results in a slight increase in hydrolytic activity. Nevertheless, there is approximately a 2-fold activity increase in polymeric liposomes compared to the monomer which can be

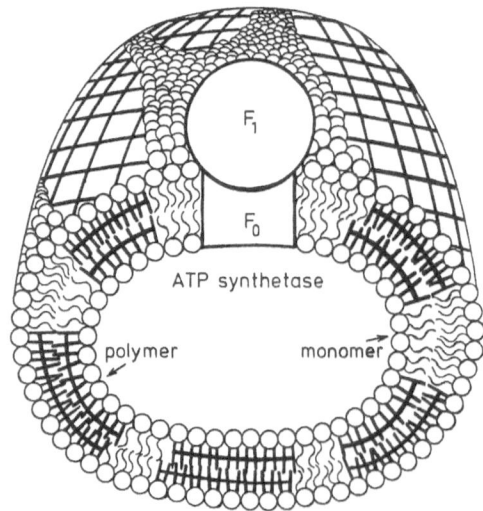

Fig. 42. Schematic presentation of ATP synthetase incorporated into a partially polymerized liposome [24]

ascribed to a structural change in the bilayer organization during polymerization. DSC data indicate residual "monomeric domains" in the polymerized liposomes. The ATPase is most probably embedded in these domains which are stabilized by the polymer matrix (Fig. 42). These polymeric proteoliposomes have a considerably higher long-term stability and activity than ATPase containing soybean lecithin liposomes.

4.2.4 Surface Recognition of Polymeric Liposomes

Biological membranes are not only responsible for compartmentation of living matter but are also able to recognize each other, which for instance, prevents unlimited tissue growth. Biomembranes are also capable of recognizing target proteins such as antibodies resulting i.e. in cell agglutination. These recognizing processes are performed by hydrophilic head groups of special lipid and protein components in the membrane.

In order to imitate biological surface recognition with polymeric membrane systems recognizable natural lipids can either be incorporated into polymerizable membranes or polymerizable lipids carrying recognizable head groups have to be synthesized. For first simple recognition experiments sugars were chosen as recognizable head groups. Sugars are recognized by lectins (plant proteins with specific binding sites for certain carbohydrates). The most frequently used lectin, Concanavalin A (Con A) (Fig. 43) is able to bind α-mannopyranoside and α-glucopyranoside. It is a tetramer, each subunit carrying one binding site. If a low molecular weight sugar is added to an aqueous lectin solution, the binding sites of the protein are saturated without a

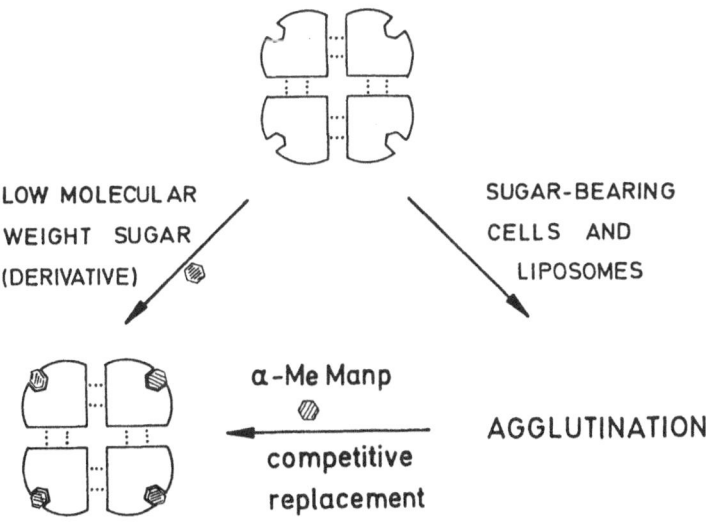

Con A (α-Manp-, α-Glcp-specific) – TETRAMER

Fig. 43. Scheme of lectin (Concanavalin A) — sugar interactions

turbidity increase being observed. If, however, the sugar moiety is part of an aggregate such as a liposome or a cell, these aggregates are agglutinated and precipitation occurs [73]. Adding a low molecular weight sugar to these aggregates will result in a competitive replacement of the aggregates by the low molecular weight sugar. The precipitate will redissolve with formation of the original liposome or cell dispersion and saturated lectin.

Based on these principles, glycolipids with different sugar head groups were synthesized carrying polymerizable diacetylene groups in the hydrophobic chain (Fig. 44) [74]. When an aqueous dispersion of (48), obtained by ultrasonication, is polymerized by UV irradiation the usual color change from blue to red takes place. In contrast to this, no liposomes are formed by (49), which is most probably due to the strong hydrogen bonding tendency of the hydrazide linkage making (49) excessively rigid and poorly dispersible in water.

Interaction of (48) with Con A has been studied in monolayers and liposomes. After injecting a Con A solution under a monolayer of (48) held under constant surface pressure, a considerable expansion of the film can be observed owing to the strong interaction of the lectin with the glycolipid. After addition of Con A to a polymerized liposomal solution of (48), agglutination occurs within a few seconds, and a red precipitate is formed leaving a colorless supernatant (Fig. 45). On addition of α-methylmannopyranoside the red precipitate is redispered yielding a clear red solution of the original liposomes. This process can be monitored by measuring the increase and decrease in turbidity (absorption at 360 nm) as a function of time. Figure 46 shows that the equilibrium is reached within approximately one minute and that the agglutination process is completely reversible. This agglutiantion does not occur with liposomes not carrying sugars at their surface. Presently attempts are being made to incorporate lectins into polymerized liposomal membranes in order to mimic vesicle-vesicle recognition between lectin and sugar bearing liposomes. The possibility of incorporating membrane proteins into polymerized membrane systems has already been demonstrated with the F_0-F_1-ATP synthetase complex (see 4.2.3).

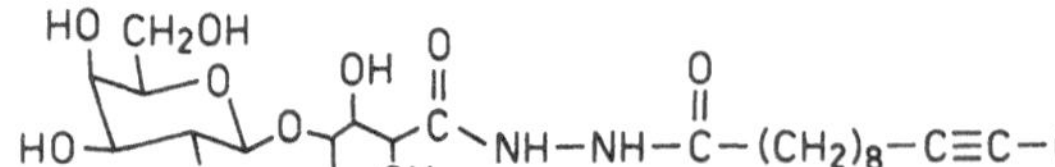

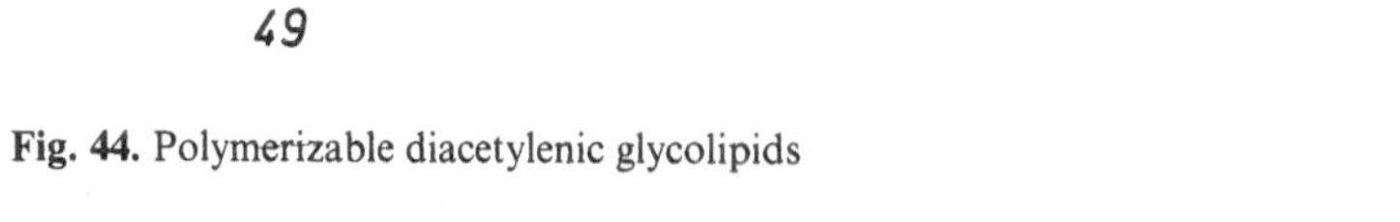

Fig. 44. Polymerizable diacetylenic glycolipids

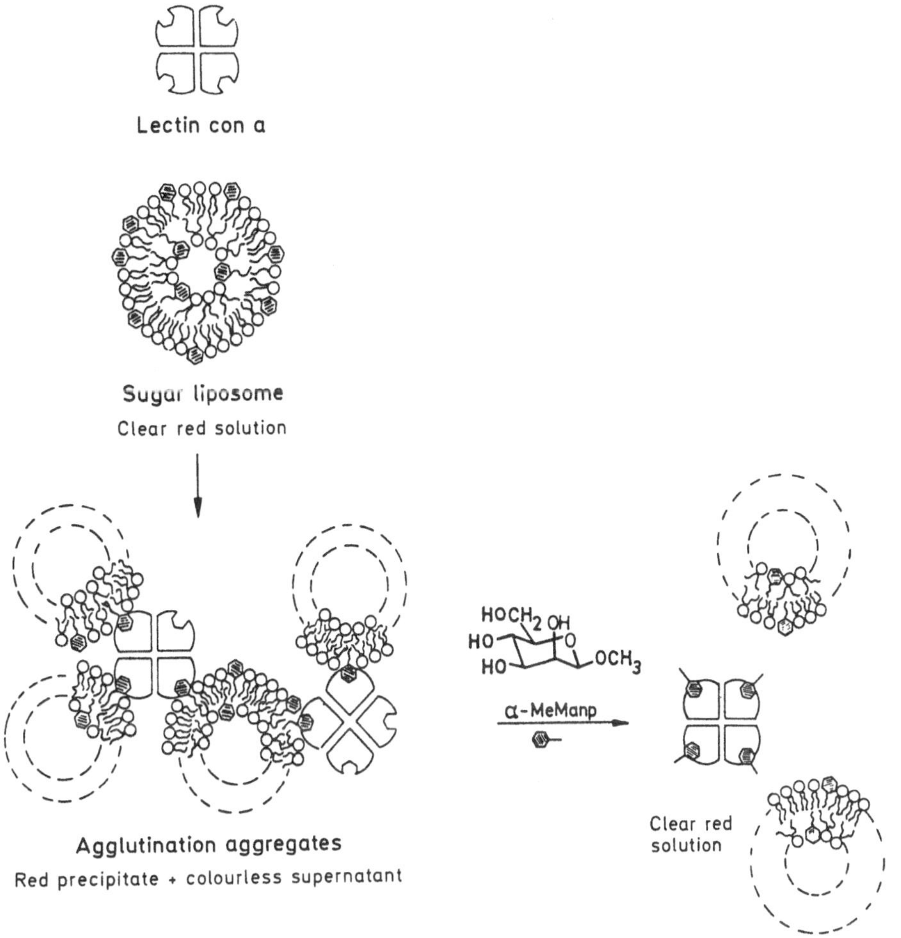

Lectin con a
Sugar liposome
Clear red solution
HOCH2 OH
HO
O
HO
OCH3
α-MeManp
Agglutination aggregates
Red precipitate + colourless supernatant
Clear red
solution

Fig. 45. Scheme of the interaction between Concanavalin A and sugar-bearing liposomes

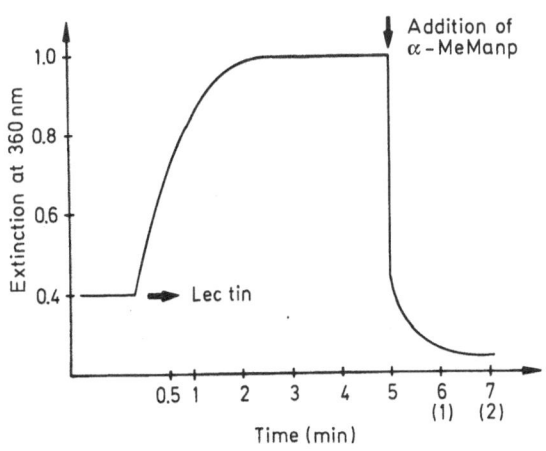

Addition of
α-MeManp
1.0
0.8
0.6
0.4
Extinction at 360 nm
Lec tin
0.5 1 2 3 4 5 6 7
(1) (2)
Time (min)

Fig. 46. Effect of addition of lectin Concanavalin A and α-methyl-mannopyranoside to vesicles of (*48*), measured by change of absorption at 360 nm

4.3 Biomembranes with Polymeric Accessories

In contrast to the synthetic route to polymeric biomembrane models (which involves putting together all kinds of natural and synthetic membrane forming substances) the method discussed in this chapter uses intact natural cell membranes and modifies them by incorporating polymerizable "accessories". The basic procedure applied for the incorporation process is the fusion of two membrane systems, e.g. living cell and polymerizable vesicle. All fusion methods known thus far have only been applied successfully to cell–cell fusion. In order to fuse a polymerizable vesicle with a natural cell membrane, investigations have to be carried out as to whether particular fusion methods can be used for liposome–liposome fusion, before applying them to cell-liposome fusion. Among the many fusion methods known [75, 76] a very selective method recently described by Zimmermann et al. [61] will be discussed in more detail.

4.3.1 Electrical Field-Induced Membrane Fusion

In the following, a physical fusion method is described based on the electrical break-down of membranes. No attempt is made to explain this procedure in full detail (for an extensive review see [61]), but rather a qualitative picture is given in order to provide a general understanding of the process.

Essentially, this method is based on subjecting cell membranes to a short external electric field pulse of an intensity comparable to the electric field strength of the membrane [77]. Under these conditions the membrane breaks down locally and becomes permeable. This process is reversible, i.e. the membrane reconstitutes its original properties in time intervals which can be experimentally controlled.

Looking at the biomembrane model of Singer and Nicolson (Fig. 2) in terms of physics, a membrane can be represented by a parallel connection of a plate capacitor and a resistor. The external aqueous solution and the polar head groups of the lipids represent the plates of the capacitor while the hydrophobic membrane interior forms a dielectric. It is well known that capacitors can only be charged to a certain maximum voltage. Above this critical value, depending on the separation distance of the plates and the type of dielectric, electrical breakdown occurs. Electrical breakdown is associated with an extreme increase in electric conductivity of the capacitor. This process is usually irreversible, i.e. the capacitor is destroyed. In self-generating capacitors the original resistor and capacitor properties are restored. Under certain experimental conditions, biological membranes and artificial lipid bilayers behave

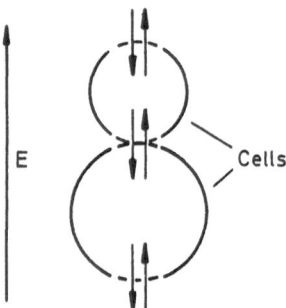

Fig. 47. Scheme of an electrical breakdown of two cells alligned in direction of the electric field [61]

analogously to these self-generating capacitors. The dielectric breakdown of a membrane can be explained in terms of local electromechanical compression. Above a certain level of compression, the membrane becomes locally unstable, possibly resulting in the formation of pores filled with electrolyte. The reversible breakdown of a cell membrane brings about a perturbation of the bilayer, permitting material exchange between the cell and its environment. If two cells adhering to each other are oriented in the external field direction, as illustrated in Fig. 47, dielectric breakdown occurs at the poles of the cells and in the zone of contact between the two cells. An exchange of substances between the cells and their evironment and an intracellular mass transport are both possible under these conditions. If the membrane contact is close enough (1–2 nm) lipid molecules are able to diffuse from one membrane to the other (Fig. 48 b). During the resealing process lipid bridges may form between the two cells, i. e. the two cells do not reseal separately in the contact zone (c). These bridges bring about very small radii of curvature (d) and in turn, high surface tension. This is why the next step, the fusion of the two cells into one sphere is energetically favored. This means that the entire electrically induced fusion process is based exclusively on physical techniques consisting of two steps: first, the membranes of cells are brought into close contact and second, in the resulting cell aggregates, fusion is induced by dielectric breakdown.

How can cells be brought into the proper aligment required for electrically induced fusion? When cells are placed in an inhomogeneous electric field, as demonstrated in Fig. 49, they start to migrate in the direction of larger field strength. This is caused

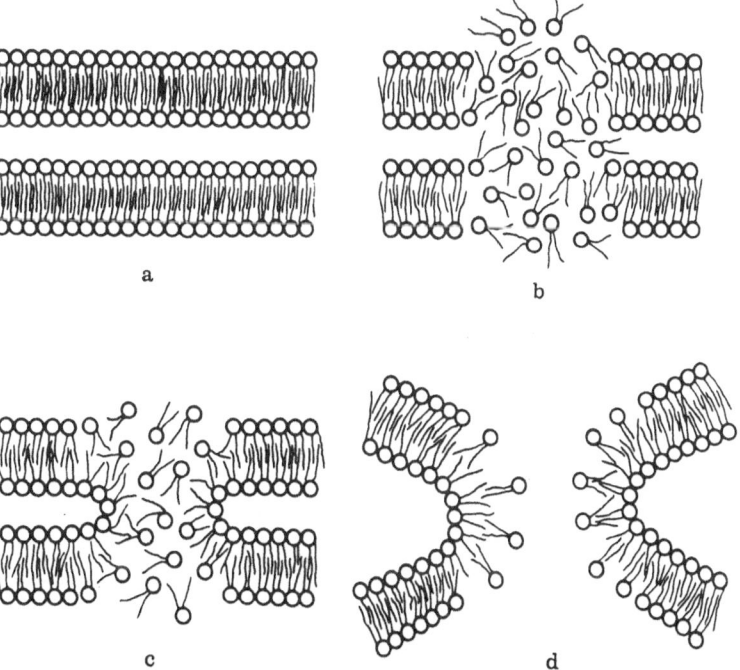

a

b

c

d

Fig. 48a–d. Model for the molecular processes which may occur during electrically induced fusion. see also text for details [61]

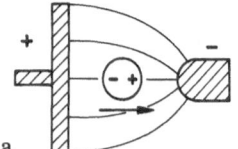 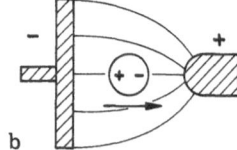

Fig. 49a and b. Diagrammatic representation of a neutral particle in an inhomogeneous electric field. **a)** the particles are able to migrate, because the field intensity is not equal on both sides resulting in a net force; so-called dielectrophoresis. **b)** The direction of dielectrophoresis is independent of the polarity of the field. In an alternating field all particles will move in the direction of higher field strengths [61]

by the different field strengths on the two sides of the dipole induced in the cell. The result is a net force pulling the particle towards larger field strength (a). This effect is known as dielectrophoresis [79]. The direction of the net force exerted on the cell does not change when the direction of the field is reversed (b). However, applying an alternating external field masks the possible net charge of the cell so that only dielectrically induced migration towards higher field intensity can occur. When cells approach each other during their migration along the field lines they attract each other because of their dipole moments (Fig. 50), therefore arranging themselves like "strings of pearls" along the field lines. This configuration remains stable as long as the alternating external field is applied. If the field is removed the chains break up because the cells repel each other due to their net charge and Brownian motion. This means that dielectrophoresis is a reversible process allowing membranes to come into close contact.

Electric field-induced fusion has been applied to a vast variety of cells including human erythrocytes and liposomes made from asolectin and egg phosphatidylcholine. To what extent this method can be utilized for fusing polymerizable vesicles will be demonstrated in the following.

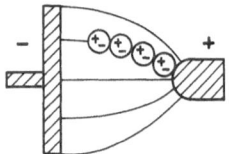

Fig. 50. Diagrammatic representation of "mutual dielectrophoresis" [61]. If the particles approach during dielectrophoresis they are attracted to each other due to their dipoles. This leads to the formation of "pearl chains" of cells

4.3.2 Liposome-Liposome Fusion

The preparation of giant liposomes [79] made it possible to apply electrical field-induced fusion to the fusion of vesicles made from natural and polymerizable lipids. Since in dielectrophoresis the net force pulling vesicles towards higher field intensities is proportional to the volume of the particle, only large vesicles can be aligned in parallel to the field lines between the electrodes.

In a general preparation technique for giant liposomes the lipid (3 mg) is dissolved in 1 ml chloroform/methanol 10:1 and the solvent is evaporated in a horizontally rotating cylindrical reaction vessel. During this process the lipid adheres to the vessel walls forming a thin, homogeneous film. Distilled water (5–10 ml) or salt solution (<5 mM) is added, followed by incubation for 1–4 h at 70 °C without agitation. During this period the lipid film detaches from the vessel wall and forms vesicles larger than 1 μm in diameter.

Using the large vesicles, the fusion process can be monitored in a phase contrast microscope. In different fusion techniques [80–82] very small, submicroscopic vesicles (<100 nm) were used and fusion could only be followed by indirect methods.

Natural lipids used for fusion experiments were mainly phospholipids with different chain lengths and their mixtures with cholesterol. As polymerizable lipids, butadienic derivatives with a phosphatidylcholine (19) and a dimethylammonium head group (26) were used in the fusion experiments.

In Fig. 51 polymerizable vesicles made from butadienic lipid (26) and cholesterol (1:1 mixture) are oriented between electrodes applying a field strength of 2–4 kV/cm. The diameter of the two large vesicles is about 40 μm [83]. For establishing an optimum membrane contact, the field strength is slightly increased which sometimes causes flattening of the vesicles in the contact zone (Fig. 51a). It is known from cell–cell fusion experiments that sufficiently close membrane contact is necessary for fusion. On the other hand, excessively high field strengths lead only to an elongation

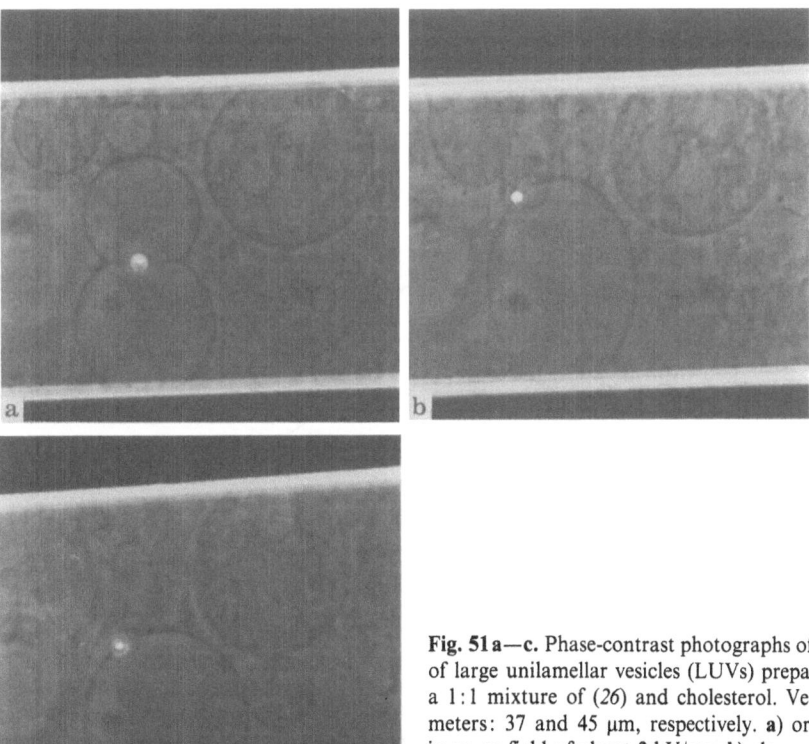

Fig. 51 a—c. Phase-contrast photographs of a fusion of large unilamellar vesicles (LUVs) prepared from a 1:1 mixture of (26) and cholesterol. Vesicle diameters: 37 and 45 μm, respectively. a) orientation in an ac field of about 2 kV/cm; b) elongated fused liposome, one second after application of a 30 μs, 140 V/cm field pulse; c) spherical new vesicle after turning off ac field; new diameter: 51 μm [83]

of the aggregated liposomes or cells. Fusion is initiated by a field pulse of 30–90 kV/cm and 20–50 μs duration. The intermingling of the membranes occurs within a fraction of a second. The membrane boundary between the two vesicles disappears forming an oval (Fig. 51 b). A completely spherical liposome is formed after turning off the field (Fig. 51 c). If the vesicles are unilamellar the whole fusion process is completed within one second. In the case of multilamellar vesicles fusion time is extended to about 3–6 seconds. Fusion times for cells are usually in the range of minutes [61].

During the fusion process the relative surface area decreases with increasing volume indicating a loss of membrane material (about 22 % in Fig. 51). In analogy to the fusion process of protoplasts it can be assumed that the excess lipid is removed in form of small, submicroscopic vesicles (Fig. 52). The electric breakdown in the membrane contact zone leads to the formation of several pores in which lipid molecules are randomly oriented (Fig. 52 b). The molecules reorient forming submicroscopic vesicles and the new membrane of the fused vesicle (Fig. 52 c). Thus, fused giant liposomes should contain small, submicroscopic vesicles. This could possibly be proven by using fluorescence-labelled lipids for liposome fusion.

In preliminaty experiments, problems occured during the fusion of giant liposomes with cells. The induced dipole of cells is much larger than that of vesicles leading primarily to cell–cell contacts and thus to cell–cell fusion. By using vesicles filled with electrolyte or multilamellar vesicle dispersions this problem could be overcome.

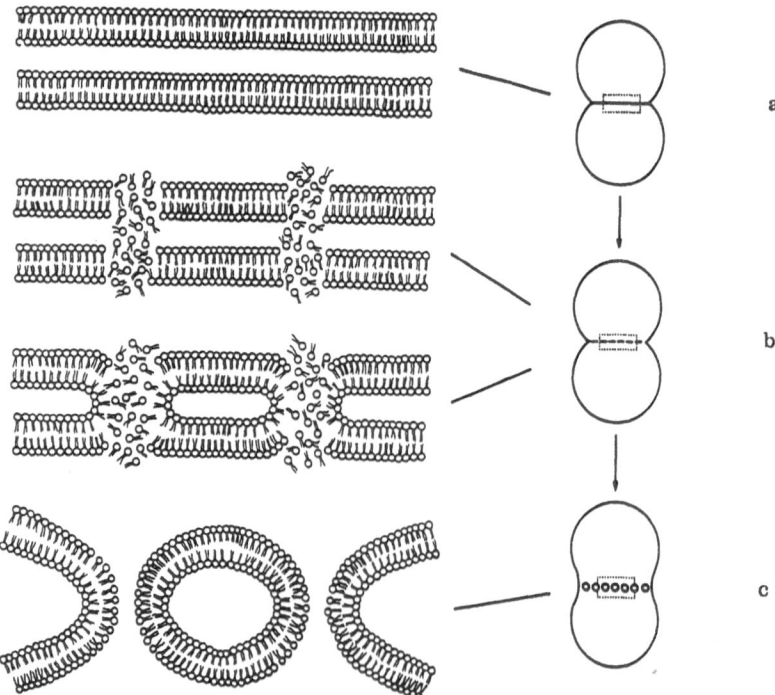

Fig. 52a–c. Scheme of the fusion process of giant liposomes and the formation of small unilamellar vesicles (SUV) at the interface. **a)** lipid bilayers in contact; **b)** pores generated by electric breakdown and lipid reorientation forming SUVs; **c)** reconstitution of lipid membranes: formation of a fused giant liposome and SUVs [83].

5 Concluding Remarks

In this contribution, it was intended to show how polymer chemistry is able to provide stabilized model membrane systems which, in addition, exhibit biological properties. Thus far, however, these systems have solely been characterized in vitro. In order to find out how good a biomembrane model really is, its performance has to be tested in the natural environment of cells, i.e. in vivo. In the future, emphasis will have to be placed upon studies on the fate of polymeric membrane models in, e.g., the blood circulation. How fast are they cleared out of the blood stream by the reticulo-endothelial system or how well do they retain entrapped substances (drugs, enzymes, markers) compared to their nonpolymerized counterparts? Does surface recognition demonstrated in vitro still function in vivo? Can they be targeted with a "homing device" and how do they behave once they have reached their target? What happens to the polymer chain in the organism? Are polypeptide vesicles on the basis of long-chain α-amino acids indeed biodegradable? Besides these in vivo studies, which are particularly important for polymeric biomembrane models prepared via the synthetic route, an even more interesting subject to study is the biological behavior of natural cells with "polymeric accessories". Are these cells — once their membrane has been modified with polymeric components — still able to live? Would they pass on their modification in mitosis?

The list of arbitrarily collected and open questions is endless. The studies carried out on polymeric membrane models discussed above can only be regarded as a first hesitant step to bridge the gap between polymer chemistry and life science.

6 Appendix

More than a year has elapsed since the first submission of our manuscript and research in the field of polymeric membranes has continued to advance. Major progress has been made in all the areas mentioned above and new results and conclusions have found their way into the current literature.

Reviews have already been published by J. H. Fendler on "Polymerized Surfactant Vesicles" [91,92,93] which refer to "Novel Membrane Mimetic Systems", synthetic strategies leading to them and their characterization and potential utilization in various areas such as solar energy conversion and reactivity control. It is the intend of this appendix to bring the reader up to date on the "state of the art" of polymerized liposomes.

6.1 New Polymerizable Lipids

A series of new polymerizable amphiphiles reported recently are compiled in Table 3.

J. H. Fendler [94] has enlarged the number of polymerizable amphiphiles with the synthesis of a variety of novel surfactants and also introduced the styrene moiety (51) as another useful polymerizable unit [12]. Small unilamellar vesicles from mixtures of DPPC with (8) maintained their morphology after polymerization for months,

Table 3. Novel polymerizable lipids and surfactants

Surfactant		Ref.

$CH_3(CH_2)_{14}COOCH_2CH_2$
$CH_3(CH_2)_{14}COOCH_2CH_2$ — $N^{\oplus}$ — CH_3 / CH_2—⬡—$CH=CH_2$ $Cl^{\ominus}$ **51**   94,107)

$CH_3(CH_2)_{15}$
$H_2C=CH(CH_2)_8CONH(CH_2)_6$ — $N^{\oplus}$ — CH_3 / CH_3 $Br^{\ominus}$ **52**   94)

$CH_3(CH_2)_{14}COOCH_2CH_2$
$CH_3(CH_2)_{14}COOCH_2CH_2$ — $N^{\oplus}$ — CH_3 / CH_2—$CH=CH_2$ $Br^{\ominus}$ **53**   94)

$I^{\ominus}C\equiv N^{\oplus}$ — $\overset{\displaystyle CH_3}{\underset{}{CH}}$ — $COO(CH_2)_{11}$
$CH_3(CH_2)_{15}$ — $N^{\oplus}$ — CH_3 / CH_3 $Br^{\ominus}$ **54**   96)

$CH_3(CH_2)_{11}$—$C\equiv C$—$C\equiv C$—$(CH_2)_9$—$\overset{\displaystyle O}{\underset{\displaystyle OH}{P}}$—$OH$ **55**   97)

$HS-(CH_2)_{10}COOCH_2$
$HS-(CH_2)_{10}COOCH$
CH_2—O—$\overset{\displaystyle O}{\underset{\displaystyle O^{\ominus}}{P}}$—$OCH_2CH_2$—$\overset{\oplus}{N}$—$CH_3$ (CH_3, CH_3)   **56**   98)

$CH_3(CH_2)_{14}COOCH_2$
$CH_3(CH_2)_{14}COOCH$
CH_2—O—$\overset{\displaystyle O}{\underset{\displaystyle O^{\ominus}}{P}}$—$O$—$CH_2CH_2$—$\overset{\oplus}{N}$—$CH_2CH_2OOC$—$\overset{\displaystyle CH_3}{\underset{}{C}}=CH_2$ (CH_3, CH_3)   **57**   99)

Table 3 (continued)

Surfactant		Ref.

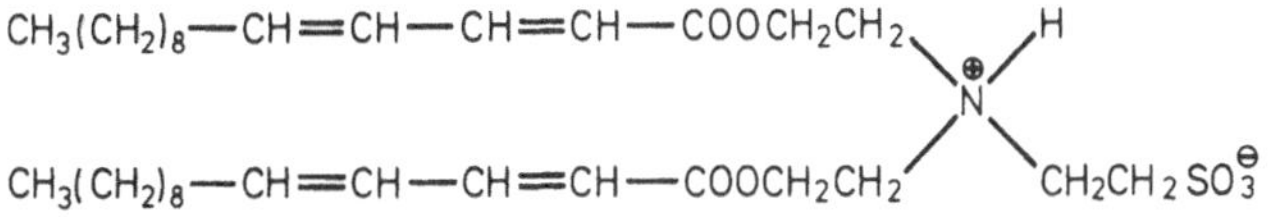

$CH_3(CH_2)_8$—CH=CH—CH=CH—$COOCH_2CH_2$, ...H

$CH_3(CH_2)_8$—CH=CH—CH=CH—$COOCH_2CH_2$' $N^{\oplus}$ $CH_2CH_2SO_3^{\ominus}$
    58 112)

$C_8F_{17}CH_2COOCH_2CH_2$, ...CH_3

$C_8F_{17}CH_2COOCH_2CH_2$' $N^{\oplus}$ CH_2—CH=CH_2 $Br^{\ominus}$
    59 101)

$C_7F_{15}CH_2NHCO$,

    C=CH_2

$C_7F_{15}CH_2NHCOCH_2'$
    60 101)

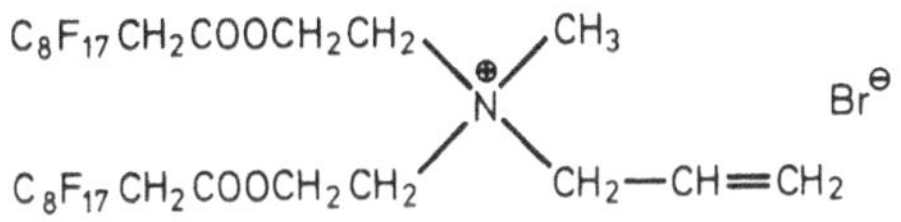

$H(CF_2)_{10}CH_2OOC$—CH=CH—CH=CH—$COOCH_2CH_2$, ...H

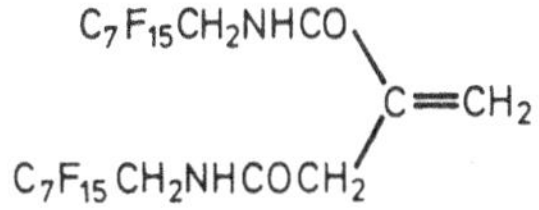

$H(CF_2)_{10}CH_2OOC$—CH=CH—CH=CH—$COOCH_2CH_2$' $N^{\oplus}$ $CH_2CH_2SO_3^{\ominus}$
    61 101)

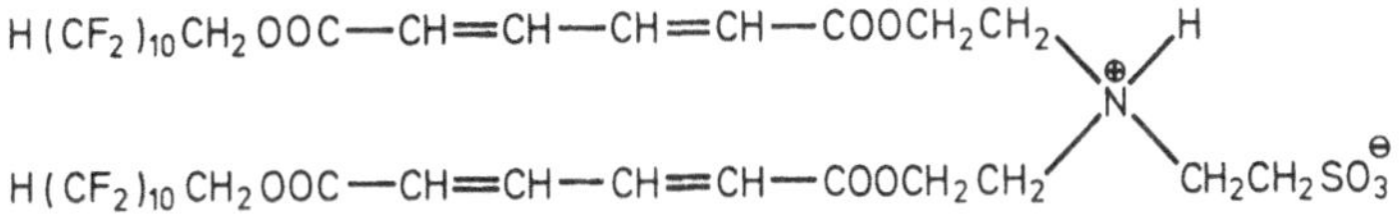

Cleavable lipids (disulfides)		Ref.

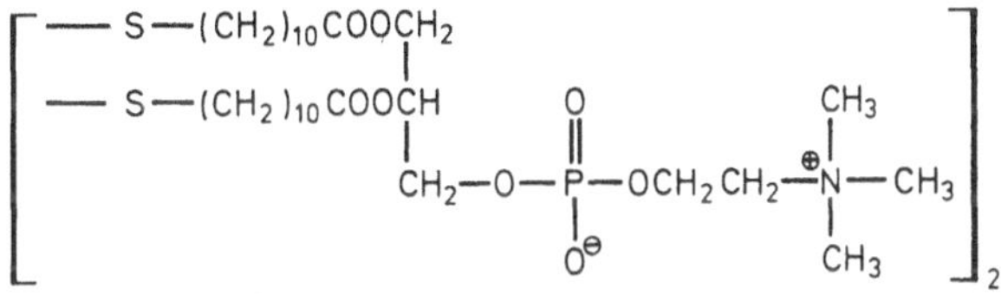

$\begin{bmatrix} --S-(CH_2)_{10}COOCH_2 \\ --S-(CH_2)_{10}COOCH \\ \qquad\qquad CH_2-O-\overset{O}{\underset{O^{\ominus}}{P}}-OCH_2CH_2-\overset{CH_3}{\underset{CH_3}{N^{\oplus}}}-CH_3 \end{bmatrix}_2$
    62 98)

$\begin{bmatrix} C_8F_{17}CH_2CONH-\underset{COOH}{CH}-S- \end{bmatrix}_2$
    63 113)

$\begin{bmatrix} C_{13}H_{27}CONH-\underset{COOH}{CH}-S- \end{bmatrix}_2$
    64 113)

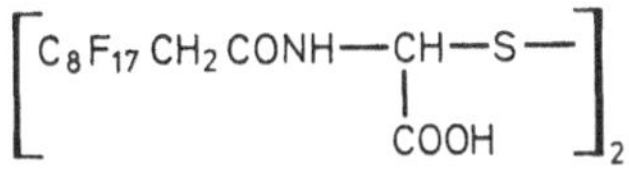

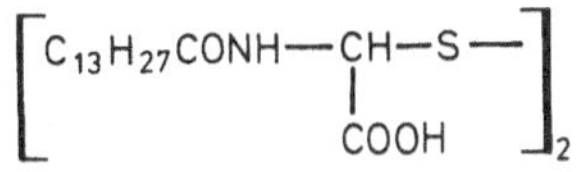

$\begin{bmatrix} C_{14}H_{29}\underset{COOH}{CH}-S- \end{bmatrix}_2$
    65 113)

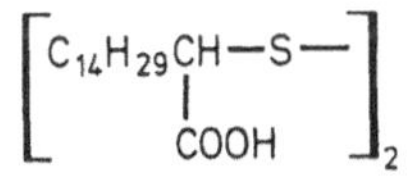

$\begin{bmatrix} HOOC(CH_2)_{11}-S- \end{bmatrix}_2$
    66 113)

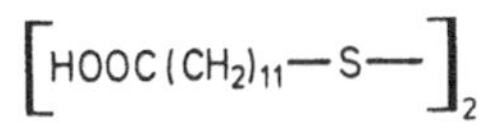

whereas nonpolymerized vesicles underwent spontaneous fusion resulting in a nonhomogeneous liposome population [95]. M. F. Roks [96] synthesized a polymerizable amphiphile (54) carrying an isocyano group at the end of one chain. This surfactant forms vesicles in aqueous dispersions and polymerizes upon addition of Ni-capronate. According to the authors' interpretation of freeze-fracture electron micrographs, crosslinks are present in the polymerized vesicles between the two halves of the bilayer. The polymerized vesicles also exhibit improved stability towards lysis by alcohol.

A diacetylene-containing chemically resistant phosphonic acid amphiphile (55) was introduced by B. Ostermeyer [97] who studied the spreading behavior of this compound at the gas/water-interface.

Reversibly polymerizable vesicles were reported by S. L. Regen [98] using a dithiol-phosphatidylcholine analogue (56, 62). The reversibility of the polymerization derives from a thiol-disulfide redox cycle. Thus, polymerization (oxidation) can be induced by UV-light or by the addition of H_2O_2, depolymerization (reduction) by addition of dithiothreitol. The vesicle structure was preserved throughout the entire reaction cycle. The polymeric vesicles show increased stability towards detergent (SDS). Reversibly polymerizable vesicles appear to be especially promising as time-release carriers of drugs.

6.2 Characterization of Polymeric Liposomes

Several investigations dealing with the characterization of polymeric membrane systems have been reported lately. A. Kusumi [99] studied phase transitions, fluidity and polarity properties of polymerized and nonpolymerized methacryloyl-derivatized phosphatidylcholine (57) vesicles by DSC and electron spin resonance (ESR).

Miscibility of a natural lipid (DMPC) and the monomeric and polymeric lecithin analogue (26) was studied in large unilamellar vesicles using freeze-fracture electron microscopy and photobleaching by H. Gaub [100]. Before polymerization the two lipids appear miscible at all compositions in the fluid state and at DMPC concentrations at or below 50 mol% in the solid state. After polymerization a two-dimensional solution of the polymer in DMPC is obtained at $T > T_g^*$ (T_g^* phase transition temperature of polymeric 26) while lateral phase segregation into DMPC-rich domains and patches of the polymer is observed $T < T_g^*$. The diameter of the polymerized lipid domains was found to average 400 Å.

Phase-separated monolayers and liposomes were characterized by R. Elbert [101] who synthesized saturated and polymerizable fluorocarbon amphiphiles (59, 60, 61) and investigated their mixing behavior with CH_2-analogues and natural lipids. In these systems the fluorocarbon compounds are incompatible with hydrocarbon lipids in a wide range of compositions and tend to form domains of pure fluorocarbon and hydrocarbon amphiphiles. The domains can be visualized by freeze-fracture electron microscopy.

The physical properties of diacetylene-containing phospholipid liposomes (25) have been investigated by D. Chapman [102,103,104] by means of UV/VIS and ^{13}C-NMR spectroscopy. They found that short linear segments of polymer are interconnected through the glycerol backbone of the lipid in identical-chain phosphatidyl cholines, but the molecular weights of the polymers are still unknown.

Molecular weight determinations have also been an area of intense current interest. The molecular weights of polymers of methacryloyl lipids have been reported by K. Dorn [105] and D. Bolikal [106]. Data of polymers of styrene-containing lipids have been determined by W. Reed [107]. The authors investigated the kinetics and mechanism of the photopolymerization of (51) and determined the M_w to be 1.0×10^8 g/mol. They also proved that the rate of monomer disappearance is considerably faster in vesicles than in ethanol, i.e. enhancement of reactivity due to the orientation of polymerizable units in liposomal membranes compared to statistical disordering in isotropic solution. D. Bolikal [106] gives data for the degree of polymerization of the methacryloyl compound (4) in a vesicle membrane. Polymerization was induced e.g. by AIBN initiation. The apparent molecular weights as determined by polystyrene calibrated GPC of freeze-dried polymeric lipid samples were M_w: 3.7×10^6; M_n: 3.98×10^5, M_w/M_n: 9.3 with an average number of monomeric units of 677.

K. Dorn [105] polymerized dialkylammonium lipids with the polymerizable methacryloyl moiety either in the head group (29) or at the end of one of the hydrophobic chains (5). GPC revealed M_w: 1.9×10^6, M_n: 3.5×10^5, M_w/M_n: 5.4 for (29) and M_w: 1.9×10^6, M_n: 3.9×10^5, M_w/M_n: 2.4 for (5). It was also found that M_w varies inversely with the time of sonication, i.e. in smaller liposomes lower-molecular-weight polymers are formed. In a following paper, K. Dorn [108] present data for the permeability of monomeric and polymeric vesicles from (29).

6.3 "Liposomes in a Net"

Principles to stabilize lipid bilayers by polymerization have been outlined schematically in Fig. 4a–d. Mother Nature — unfamiliar with the radically initiated polymerization of unsaturated compounds — uses other methods to stabilize biomembranes. Polypeptides and polysaccharide derivatives act as a type of net which supports the biomembrane. Typical examples are spectrin, located at the inner surface of the erythrocyte membrane, clathrin, which is the major constituent of the coat structure in coated vesicles, and murein (peptidoglycan) a macromolecule coating the bacterial membrane as a component of the cell wall. Is it possible to mimic Nature and stabilize synthetic lipid bilayers by coating the liposome with a polymeric network without any covalent linkage between the vesicle and the polymer? One can imagine different ways for the coating of liposomes with a polymer. This is illustrated below in Fig. 53.

There are four different routes to achieve this goal:
1. adsorption of polymer to the liposomal surface by hydrophilic and/or ionic interactions [109]
2. attachment of polymers having hydrophobic substituents by insertion of these anchor groups into the lipid core of the liposomal membrane [110]
3. fixation of polymerizable water-soluble monomers to the vesicle surface via salt formation, followed by an initiated or spontaneous polymerization
4. preparation of liposomes from amphiphiles where the polymerizable group is linked to the polar headgroup via a cleavable spacer with cleavage of the spacer after polymerization.

J. Sunamoto [110] reported a coating of the outer liposome leaflet with partially palmitoyl-derivatized polysaccharides according to route 2 in Fig. 53. These "arti-

SYNTHETIC COATED VESICLES: LIPOSOMES IN A "BASKET" (SCHEMATIC)

COATING OF LIPOSOMES BY A NON-COVALENTLY LINKED POLYMERIC NETWORK

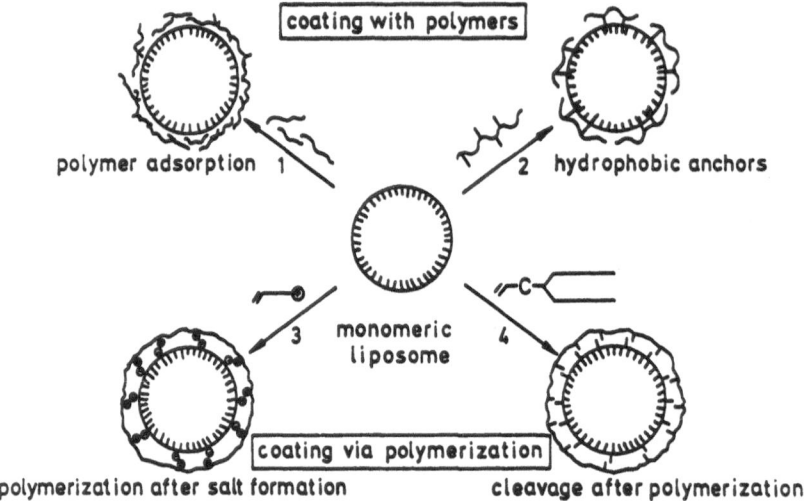

Fig. 53. Schematic representation of possible routes to stabilize liposomes via surface coating with polymers

ficial cell-wall"-like hydrophobized polysaccharides are firmly attached to the liposomes from egg lecithin as proven by gel filtration. The coated liposomes show reduced permeability for the marker 6-carboxyfluorescein and improved stability against enzymatic degradation by phospholipase D. The tissue distribution of one defined composition of polysaccharide-coated liposomes also differed remarkably from their conventional counterparts after i.v. injection into test animals showing high accumulation in spleen and lung.

Following route 3, S. L. Regen [111] built "polymer-encased vesicles" by the photo-polymerization of DODAM (dioctadecyldimethylammonium methacrylate)

$$CH_3(CH_2)_{17} \underset{CH_3(CH_2)_{17}}{\overset{+}{\diagdown}} N \underset{CH_3}{\overset{CH_3}{\diagup}} \qquad ^-OOC-\underset{\overset{|}{C}}{\overset{CH_3}{C}}=CH_2$$

Extracted polymer material had an IR spectrum identical with an authentic sample of poly(methacrylic acid). They also report data on M_w (85.000 in 0.002 M HCl) and tacticity of the poly(methacrylic acid) obtained. On the basis of Kunitake's fundamental investigation of vesicle forming dialkylammonium salts [14] similar results were reported recently by J. E. Brady [112].

B. Schlarb [113] took advantage of the spontaneous polymerization of 4-vinylpyridine (4-VP) upon addition of protic acids. Polymerization was induced when 4-VP was added to preparations of acidic dicetylphosphate (DCP) liposomes and could be monitored by UV spectroscopy. By comparison of the UV and ^{1}H-NMR spectra of the polymer with an authentic sample of poly(1,4-pyridiniumdiethylene trifluoro-acetate), it could be concluded that the polymer formed in the above reaction was the poly-1,4-species instead of the 1,2-polymer which is formed on polymerization in concentrated solution. The polymer is tightly attached to the liposomal surface and does not separate from it as shown by GPC. In transmission electron micrographs the poly(vinylpyridin) appears like threads or filaments lying on the surface of the spherically shaped liposomes. The authors therefore call this type of polymer-encased vesicles "liposomes in a net".

6.4 "Cork-screws for Corked Liposomes"

In addition to enzymatic hydrolysis of natural lipids in polymeric membranes as discussed in chapter 4.2.2., other methods have been applied to trigger the release of vesicle-entrapped compounds as depicted in Fig. 37. Based on the investigations of phase-separated and only partially polymerized mixed liposomes [101], methods to "uncork" polymeric vesicles have been developed. One specific approach makes use of cleavable lipids such as the cystine derivative (*63*). From this fluorocarbon lipid mixed liposomes with the polymerizable dienoic acid-containing sulfolipid (*58*) were prepared in a molar ratio of 1:9 [101,115]. After polymerization of the matrix forming sulfolipids, stable spherically shaped vesicles are obtained as demonstrated in Fig. 54 by scanning electron microscopy [114].

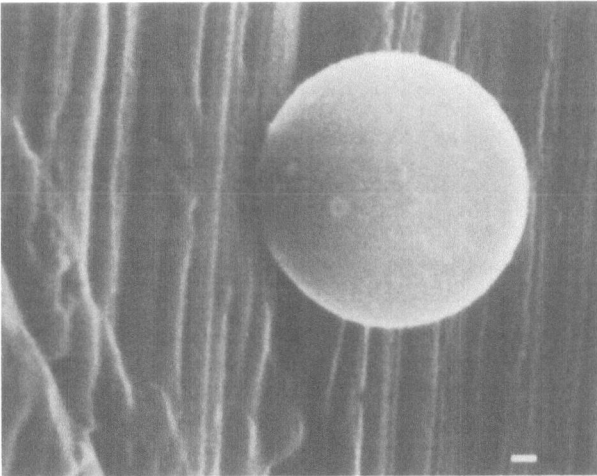

Fig. 54. Scanning electron micrograph of a polymeric liposomes from the taurine derivative *58* (90%) with phase-separated areas of the fluorocarbon amphiphile *63* (10%). Angle 60°, bar represents. 1 μm

In such phase-separated partially polymerized liposomes the "corks" consisting of cleavable lipids can be visualized by freeze-fracture electron microscopy [101] and photobleaching [100].

Possibilities to cleave the disulfide linkage of the fluorocarbon cystine leading to partially watersoluble cysteine derivatives are given in the following reaction scheme:

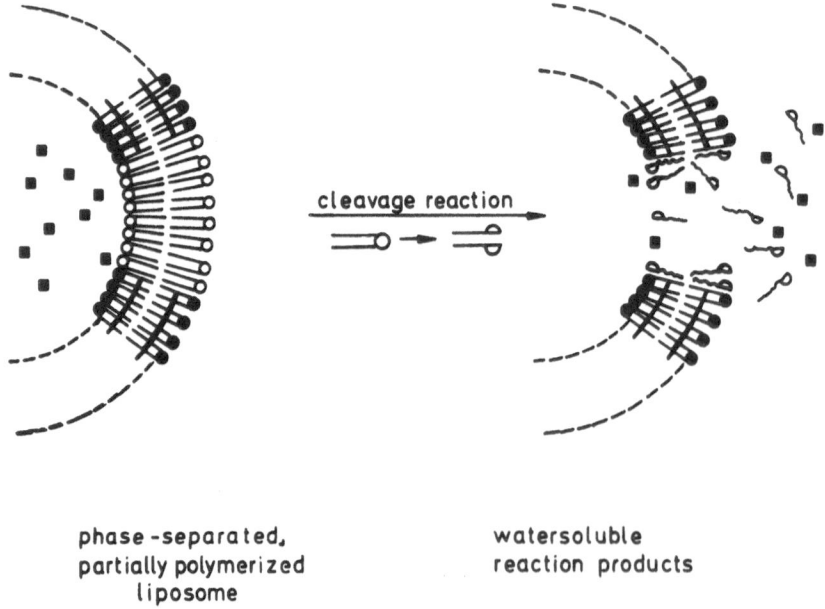

$$C_8F_{17}-CH_2-CO-NH-\underset{\underset{S}{\overset{|}{CH_2}}}{CH}-COOH$$

$$\xrightarrow[\substack{\text{or} \\ CH_2-CH-CH-CH_2 \\ SH \ OH \ OH \ SH}]{S_2O_4{}^{2\ominus}} \quad 2 \ C_8F_{17}-CH_2-CO-NH-\underset{\underset{SH}{\overset{|}{CH_2}}}{CH}-COOH$$

$$C_8F_{17}-CH_2-CO-NH-\underset{S}{CH}-COOH$$

$$\xrightarrow[pH > 7]{SO_3{}^{2\ominus}}$$

$$C_8F_{17}-CH_2-CO-NH-\underset{\underset{CH_2}{\overset{SH}{|}}}{CH}-COOH$$

$$+$$

$$C_8F_{17}-CH_2-CO-NH-\underset{\underset{SO_3^{\ominus} \ Na^{\oplus}}{\overset{|}{S}}}{CH}-COOH$$

This disulfide-thiol reduction process when applied to phase-separated liposomes of the composition given above is illustrated in Fig. 55.

cleavage reaction

phase-separated,   watersoluble
partially polymerized   reaction products
liposome

Fig. 55. Schematic illustration of the "uncorking" process

The hole formation in the liposomal membrane after treating the vesicles with reducing agents can be demonstrated by the fast and complete release of eosin, a fluorescent marker. Final proof that the polymeric backbone of the "uncorked" vesicles does not collapse comes from scanning electron microscopy. Fig. 56 shows the spherical structure of the liposomes with holes.

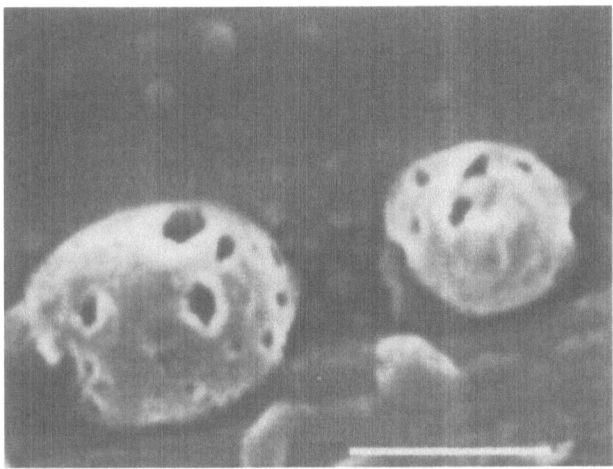

Fig. 56. Scanning electron micrograph of uncorked liposomes (bar: 1 μm)

6.5 Polymerizable Lipids and Biological Functions

Along a more biological approach, D. Chapman [116] has described the biosynthetic incorporation of diacetylene acids into the biomembranes of Acholeplasma laidlawii A, an unsaturated fatty acid auxotroph bacterium. As much as 90% of the membrane lipid acyl chains were found to consist of C_{20}-diynoic acid. Upon irradiation with UV-light, the cells and isolated membranes become coloured, due to the crosslinking of lipids by photopolymerization.

For the potential purpose of polymeric liposomes as drug carriers, two recent papers from S. L. Regen appear to be most stimulating. The authors tested the mechanical and chemical stability of vesicles from methacryloyl-containing phospholipids [117]. The stability of polymerized liposomes is significantly higher under all conditions and release of markers is monitored in the presence of serum. In another approach, functionalized polymeric liposomes were obtained [118] from a lipid with a latent aldehyde moiety in the head group and the enzyme alpha chymotrypsin was immobilized on the liposomal surface via Schiff base formation with subsequent boro-hydride reduction. The efficiency of the coupling is reported to be very high.

The incorporation of a membrane protein into a polymerizable liposome from (22) was demonstrated by R. Pabst [119]. The chromoprotein bacteriorhodopsin — a light-driven proton pump from halophilic bacteria — was incorporated into monomeric sulfolipid liposomes by ultrasonication. The resulting proteoliposomes were polymerized by UV-light. The activity of the protein was preserved and the long-term stability of the polymeric proteoliposomes was proven to be significantly superior to their soy bean lecithin counterparts. Especially from the number of innovative papers that we have had the opportunity to review in this appendix, we feel confident that the area of polymeric membranes will offer even more novel and exciting results in the future.

7 Glossary

Fig. 57. "For a start, I would strongly recommend to study this handy and easy-to-read introduction to methods of how to find liable instructions for the technique of topic selection in Life Science and Membrane Biology [90]"

Amphiphiles

Molecules consisting of a long hydrophobic part and one or two hydrophilic headgroups. Able to form micelles and/or liposomes depending on the hydrophilic-lipophilic balance (HLB).

ATP synthetase complex

Membrane-bound ATP synthetase is an enzyme present in all oxidative and/or photosynthetic organisms. It catalyses the formation of ATP from ADP and P_i using the electrochemical potential (proton gradient) across the membrane by a mechanism which is still not well understood.

Crisp's rule

Phase rule of Defay-Crisp describing the number of degrees of freedom for a system having one single plane surface (monolayer): Multi-component monolayers consisting of immiscible amphiphiles exhibit the same surface pressure for phase transitions and collapse points as the corresponding one-component monolayers, while these surface-pressures are different for mixtures of miscible amphiphiles.

Ghost cells

Cells which lost part of or all internal protein via osmotic shock techniques (e.g. hemolysis).

Hemolysis

Loss of hemoglobin by lysed red blood cells (cell membrane lysis). Controlled hemolysis (induced osmotically or electrically) yields hemoglobin-free "erythrocyte ghosts".

Langmuir-Blodgett multilayers

Adsorbed films generated by repetitively dipping a substrate into a liquid which is covered with a monolayer of the desired film material. If the material is an aliphatic organic compound with a polar end, then each dip will result in the deposition of a film whose thickness is twice the length of the compound.

Lipid bilayer

Spontaneous aggregation of lipids — double chain amphiphiles — in aqueous medium (smectic phase). Hydrophobic core of long alkyl chains (lamellar orientation) is covered by hydrophilic headgroups (see Figs. 2, 15).

Liposomes

Closed, ball-shaped structures consisting of one or more double layers of lipid molecules with aqueous interior, also called vesicles (see Fig. 17).

Micelles

Aggregates, dynamically formed from surfactants in water above a certain concentration (critical micelle concentration, cmc). In contrast to vesicles they consist of a hydrophobic core.

Phase transition temperature

In membrane systems, temperature at which the crystalline state is converted to the liquid-crystalline state (smectic phases).

Proteoliposomes

Liposomes carrying membrane proteins incorporated via different reconstitution methods (detergent dialysis, sonication etc.)

Protoplast

Mainly used to indicate living cells after digestion of the outer cell wall via lytic enzymes. The protoplasma of these cells is still enclosed by the undestroyed plasma-membrane.

Reticuloendothelial system (RES)

A widely scattered system of phagocytic cells of common ancestry fulfilling many vital functions e.g. defence against infection, antibody, blood cell, and bile pigment formation. Main components of the RES are bone marrow, spleen, liver, and lymphoid tissue.

Vesicles

Synonym for liposomes.

8 References

1. Singer, S. J., Nicolson, G. L.: Science *175*, 720 (1972)
2. Singer, S. J.: in "Cell Membranes — Biochemistry, Cell Biology, and Pathology" (Weissman, G., Clairborne, R., eds.), HP Publishing, New York, 1975, p. 44
3. Pearce, B.: Trends Biochem. Sci. Pers. Ed. *1980*, 131
4. Lehninger, A. L.: Biochemistry, Worth, New York, 1970, p. 232
5. Klein, J.: Kontakte (Merck) *3*, 29 (1980)
6. Gupta, C. M., Costello, C. E., Khorana, H. G.: Proc. Natl. Acad. Sci., U.S.A., *76*, 3139 (1979)
7. Akimoto, A., Dorn, K., Gros, L., Ringsdorf, H., Schupp, H.: Angew. Chem. Int. Ed. Engl. *20*, 90 (1981)
8. Fendler, J. H.: Acc. Chem. Res. *13*, 7 (1980)
9. Kunitake, T.: J. Macromol. Sci.-Chem. *13*, 587 (1979)
10. Baumgartner, E., Fuhrhop, J. H.: Angew. Chem. Int. Ed. Engl. *19*, 550 (1980)
11. Okahata, Y., Kunitake, T.: J. Am. Chem. Soc. *101*, 5231 (1979)
12. Tundo, P., Kippenberger, D. J., Klahn, P. L., Pietro, N. E., Tao, T. C., Fendler, J. H.: J. Am. Chem. Soc. *104*, 456 (1982)
13. Regen, S. L., Singh, A., Oehme, G., Singh, M.: J. Am. Chem. Soc. *104*, 791 (1982)
14. Regen, S. L., Czech, B., Singh, A.: J. Am. Chem. Soc. *102*, 6638 (1980)
15. Dorn, K., Klingbiel, R. T., Specht, D. P., Tyminsky, P. U., Ringsdorf, H., O'Brien, D. F.: J. Am. Chem. Soc. *106*, 1627 (1984)
16. Tundo, P., Kippenberger, D. J., Politi, M. J., Klahn, P., Fendler, J. H.: J. Am. Chem. Soc. *104*, 5352 (1982)
17. Paleos, C. M., Christias, C., Evangelatos, G. P., Dais, P.: J. Polym. Sci., Polym. Chem. Ed. *20*, 2565 (1982)
18. Bader, H., Ringsdorf, H.: J. Polym. Sci., Polym. Chem. Ed. *20*, 1623 (1982)
19. Hub, H.-H., Hupfer, B., Koch, H., Ringsdorf, H.: Angew. Chem. Int. Ed. Engl. *19*, 938 (1980)
20. Johnston, D. S., Sanghera, S., Pons, M., Chapman, D., Biochim. Biophys. Acta *602*, 57 (1980)
21. Lopez, E., O'Brien, D. F., Whitesides, T. H.: J. Am. Chem. Soc. *104*, 305 (1982)
22. Hupfer, B., Ringsdorf, H., Schupp, H.: Makromol. Chem. *182*, 247 (1981)
23. Koch, H., Ringsdorf, H.: Makromol. Chem. *182*, 255 (1981)
24. Wagner, N., Dose, K., Koch, H., Ringsdorf, H.: FEBS Lett. *132*, 313 (1981)
25. Hupfer, B., Ringsdorf, H., Schupp, H.: Chem. Phys. Lipids *33*, 355 (1983)
26. Benz, R., Praß, W., Ringsdorf, H.: Angew. Chem. Int. Ed. Engl. *21*, 368 (1982)
27. Tundo, P., Kurihara, K., Klippenberger, D. J., Politi, M., Fendler, J. H.: Angew. Chem. Int. Ed. Engl. *21*, 81 (1982)
28. Fukuda, K., Shibasaki, Y., Nakahara, H.: J. Macromol. Sci.-Chem. *15*, 999 (1981)
29. Sackmann, E.: Ber. Bunsenges. Phys. Chem. *78*, 929 (1974)
30. Gaines, G. L.: "Insoluble Monolayers at Liquid-Gas Interfaces", Interscience, New York, 1966
31. Tien, H. T.: "Bimolecular Lipid Membranes, Theory and Practice", Dekker, M., New York, 1974
32. Szoka, F., Papahadjopoulos, D.: Ann. Rev. Biophys. Bioeng. *9*, 467 (1980)
33. Gros, L., Ringsdorf, H., Schupp, H.: Angew. Chem. Int. Ed. Engl. *20*, 305 (1981)
34. Tabor, D.: J. Colloid Interface Sci. *75*, 240 (1980)
35. Langmuir, I.: J. Am. Chem. Soc. *39*, 1848 (1917)
36. Hub, H.-H., Hupfer, B., Koch, H., Ringsdorf, H.: J. Macromol. Sci-Chem. *15*, 701 (1981)
37. Cadenhead, D. A.: Recent Progr. Surface Sci. *3*, 169 (1970)

38. Dörfler, H. D., Rettig, W.: Colloid Polym. Sci. *258*, 839 (1980)
39. Gorter, E., Grendel, F.: J. Exp. Med. *41*, 439 (1925)
40. Wiedmer, T., Brodbeck, U., Zahler, P., Fulpius, B. W.: Biochim. Biophys. Acta *506*, 161 (1978)
41. Demel, R. A., Geurts van Kessel, W. S. M., Zwaal, R. F. A., Boelofsen, B., van Deenen, L. L. M.: Biochim. Biophys. Acta *406*, 97 (1975)
42. Crisp, D. J.: in "Surface Chemistry" (Proc. Joint Meet. Faraday Soc. Chem. Phys.), Butterworths, London, 1967, p. 1944
43. Singer, M. A.: Biochem. Pharmacol. *29*, 2651 (1980)
44. Blume, A.: Biochim. Biophys. Acta *557*, 32 (1979)
45. Naegele, D., Ringsdorf, H.: in "Polymerization of Organized Systems", Midland Macromolecular Monographs, Vol. 3 (Elias, H. G., ed.) Gordon & Breach, New York, 1977, p. 79
46. Hupfer, B., Ringsdorf, H.: in "Surface and Interfacial Aspects of Biomedical Polymers" (Andrade, J. D., ed.) Vol. 1, Plenum, New York, 1984, in the press
47. Day, D., Hub, H.-H., Ringsdorf, H.: Isr. J. Chem. *18*, 325 (1979)
48. Day, D., Ringsdorf, H.: Makromol. Chem. *180*, 1059 (1979)
49. Tieke, B., Lieser, G., Wegner, G.: J. Polym. Sci., Polym. Chem. Ed. *17*, 1631 (1979)
50. Wegner, G.: Makromol. Chem. *154*, 35 (1972)
51. Day, D., Lando, J. B.: Macromolecules *13*, 1478 (1980)
52. Folda, T., Gros, L., Ringsdorf, H.: Makromol. Chem. Rapid Commun. *3*, 167 (1982)
53. Crisp, D. J.: in "Surface Chemistry", Interscience, New York, 1949, p. 17
54. Bangham, A. D., Standish, M. M., Watkins, J. C.: Mol. Biol. *13*, 238 (1965)
55. Bodemann, H., Passow, H.: J. Membr. Biol. *8*, 1 (1972)
56. Huang, C.: Biochemistry *8*, 344 (1969)
57. Gebicki, J. M., Hicks, M.: Chem. Phys. Lipids *16*, 142 (1976)
58. Weinstein, J. N., Yoshikami, S., Heukart, L., Blumenthal, R., Hagins, W. A.: Science *195*, 489 (1977)
59. Ebelhäuser, R., Spiess, H. W. to be published, Makromol. Chem. Rapid Commun. *5*, 403 (1984)
60. Weismann, G., Clairborne, R. (eds.): "Cell Membranes — Biochemistry, Cell Biology, and Pathology", HP Publishing, New York, 1975
61. Zimmermann, U., Scheurich, P., Pilwat, G., Benz, R.: Angew. Chem. Int. Ed. Engl. *20*, 325 (1981)
62. Büschl, R., Hupfer, B., Ringsdorf, H.: Makromol. Chem. Rapid Commun. *3*, 589 (1982)
63. de Kruyff, B., Demel, R. A., Slotboom, A. J., van Deenen, L. L. M., Rosenthal, A. F.: Biophys. Acta *307*, 1 (1973)
64. Chapman, D., Owens, N. F., Phillips, M. C., Walker, D. A.: Biochim. Biophys. Acta *183*, 458 (1969)
65. Phillips, M. C., Joos, P.: Kolloid Z. Z. Polym. *238*, 499 (1970)
66. Patil, G. S., Cornwell, D. G.: J. Lipid Res. *18*, 1 (1977)
67. Defay, R., Prigogine, I., Bellemans, A., Everett, D. H.: in "Surface Tension and Adsorption", London, 1966, Chap. 6
68. Demel, R. A., van Deenen, L. L. M., Pethica, B. A.: Biochim. Biophys. Acta *135*, 11 (1967)
69. Gaub, H., Sackmann, E., Büschl, R., Ringsdorf, H.: Biophys. J. *45*, 725 (1984)
70. Yatvin, M. B., Kreutz, W., Herwitz, B. A., Shinitzky, M.: Science *210*, 1253 (1980)
71. Weinstein, J. N., Magin, R. L., Yatvin, M. B., Zaharko, D. S.: Science *204*, 188 (1979)
72. Kano, K., Tanaka, Y., Ogawa, T., Shimomura, M., Okahata, Y., Kunitake, T.: Chem. Lett. *1980*, 421
73. Goldstein, I. J., ed.: "Carbohydrate-Protein Interactions", ACS-Sympos. Ser., Vol. 88, Washington D. C., 1979
74. Bader, H., Ringsdorf, H., Skura, J.: Angew. Chem. Int. Ed. Engl. *20*, 91 (1981)
75. Poste, G., Nicolson, G. L.: "Cell Surface Reviews", Vol. 5, Elsevier, Amsterdam, 1976
76. Ringertz, N. R., Savage, R. E.: "Cell Hybrids", Academic, New York, 1976
77. Zimmermann, U., Schultz, J., Pilwat, G.: Biophys. J. *13*, 1005 (1973)
78. Scheurich, P., Zimmermann, U., Schnabl, H.: Plant. Physiol. *67*, 849 (1981)
79. Hub, H.-H., Zimmermann, U., Ringsdorf, H.: FEBS Lett. *140*, 254 (1982)

80. Papahadjopoulos, D., Hui, S., Vail, W. J., Poste, G.: Biochim. Biophys. Acta *448*, 245 (1976)
81. Poste, G., Nicolson, G. L.: eds. "Membrane Fusion", Elsevier Amsterdam, 1978
82. Wilschut, J., Duezguennes, N., Papahadjopoulos, D.: Biochemistry *20*, 3126 (1981)
83. Büschl, R., Ringsdorf, H., Zimmermann, U.: FEBS Lett. *150*, 38 (1982)
84. Fendler, J. H.: "Membrane Mimetic Chemistry", Wiley-Interscience, New York 1982
85. Old, L. J.: Sci. Am. *236* (5), 62 (1977)
86. Sackmann, E., Rüppel, D., Gebhardt, C.: in "Liquid Crystals of One- and Two-dimensional Order" (Helfrich, W., Heppke, G., eds.), Springer Series in Chemical Physics, Vol. 11, Springer, Berlin 1980, p. 309
87. Boheim, G., Hauke, W., Eibl, H.: Proc. Natl. Acad. Sci. U.S.A. *77*; 3403 (1980)
88. Ash, P. S., Bunce, A. S., Dawson, C. R., Hider, R. C.: Biochim. Biophys. Acta *510*, 216 (1978)
89. Dawson, C. R., Drake, A. F., Helliwell, J., Hider, R. C.: Biochim. Biophys. Acta *510*, 75 (1978)
90. von Saalfeld, H.: Bestiarium Academicum, Mainz 1983
91. Fendler, J. H.: Science *223*, 888 (1984)
92. Fendler, J. H., Tundo, P.: Acc. Chem. Res. *17*, 3 (1984)
93. Fendler, J. H.: Chem. & Eng. News *62*, 25 (1984)
94. Kippenberger, D., Rosenquist, K., Odberg, L., Tundo, P., Fendler, J. H.: J. Am. Chem. Soc. *105*, 1129 (1983)
95. Kurihara, K., Fendler, J. H.: J. Chem. Soc. Chem. Commun. *1983*, 1188
96. Roks, M. F. M., Visser, H. G. J., Zwikker, J. W., Verklej, A. J., Nolte, R. J. M.: J. Am. Chem. Soc. *105*, 4507 (1983)
97. Ostermayer, B., Vogt, W.: Makromol. Chem. Rapid Commun. *3*, 563 (1982)
98. Regen, S. L., Yamaguchi, K., Samuel, N. K. P.. Singh, M.: J. Am. Chem. Soc. *105*, 6354 (1983)
99. Kusumi, A., Singh, M., Tirrell, D. A., Oehme, G., Singh, A., Samuel, N. K. P., Hyde, J. S., Regen, S. L.: J. Am. Chem. Soc. *105*, 2975 (1983)
100. Gaub, H., Sackmann, E., Büschl, R., Ringsdorf, H.: Biophys. J. *45*, 725 (1984)
101. Elbert, R., Folda, T., Ringsdorf, H.: J. Am. Chem. Soc. in press
102. Johnston, D. S., McLean, L. R., Whittam, M. A., Clark, A. D., Chapman, D.: Biochemistry *22*, 3194 (1983)
103. Pons, M., Villaverde, C., Chapman, D.: Biochim. Biophys. Acta *730*, 306 (1983)
104. Leaver, J., Alonso, A., Durrani, A. A., Chapman, D.: Biochim. Biophys. Acta *732*, 210 (1983)
105. Dorn, K., Patton, E. V., Klingbiel, R. T., O'Brien, D. F., Ringsdorf, H.: Makromol. Chem. Rapid Commun. *4*, 513 (1983)
106. Bolikal, D., Regen, S. L.: Macromolecules *17*, 1287 (1984)
107. Reed, W., Guterman, L., Tundo, P., Fendler, J. H.: J. Am. Chem. Soc. *106*, 1897 (1984)
108. Dorn, K., Klingbiel, R. T., Specht, D. P., Tyminski, P. N., Ringsdorf, H., O'Brien, D. F.: J. Am. Chem. Soc. *106*, 1627 (1984)
109. Kunitake, T., Yamada, S.: Polym. Bull. *1*, 35 (1978)
110. Sunamoto, J., Iwamoto, K., Yuzuriha, T., Katayama, K.: Polym. Sci. Technol. (Plenum) 1983, *23* (Polym. Med.) 157
111. Regen, S. L., Shin, J. S., Yamaguchi, K.: J. Am. Chem. Soc. *106*, 2446 (1984)
112. Brady, J. E., Evans, D. F., Kachar, B., Ninham, B. W.: J. Am. Chem. Soc. *106*, 4279 (1984)
113. Aliev, K. V., Ringsdorf, H., Schlarb, B., Leister, K.-H.: Makromol. Chem. Rapid Commun. *5*, 345 (1984)
114. Büschl, R., Folda, T., Ringsdorf, H.: Makromol. Chem. Suppl. *6*, 245 (1984)
115. Folda, T., Lando, J. B., Ringsdorf, H.: in preparation
116. Leaver, J., Alonso, A., Durrani, A. A., Chapman, D.: Biochim. Biophys. Acta *727*, 327 (1983)
117. Juliano, R. L., Hsu, M. J., Regen, S. L., Singh, M.: Biochim. Biophys. Acta *770*, 109 (1984)
118. Regen, S. L., Singh, M., Samuel, N. K. P.: Biochim. Biophys. Res. Commun. *119*, 646 (1984)
119. Pabst, R., Ringsdorf, H., Koch, H., Dose, K.: FEBS Lett. *154*, 5 (1983)
120. Hupfer, B., Ringsdorf, H.: Chem. Phys. Lipids *33*, 263 (1983)

M. Gordon (Editor)
Received September 1 st, 1983

Polyamides as Barrier Materials

Hiroshi Sumitomo and Kazuhiko Hashimoto
Faculty of Agriculture, Nagoya University, Chikusa, Nagoya 464, Japan

In the performance data of various polyamide and related membranes published to date there should be valuable information for molecular design of more excellent barrier materials. But at present a means for their evaluation and optimization is still not clear. One of the reasons may at least come from the competitive flood of proposals for the detailed mechanisms of reverse osmosis, e.g. the solution-diffusion model, the sieve model, the preferential sorption model and so on. [109]*

It is important for the molecular design of a superior barrier materials to take into account the balance and distribution of hydrophilic and hydrophobic microdomains as well as the stiffness and flexibility of a macromolecule, in order to place the subject on the fundamental understanding of the interactions among solvents, solutes, and macromolecules.

1 Introduction

Many synthetic membranes are known to be useful for separation of water and various sizes of solutes from aqueous solutions by selective separation, for examples reverse osmosis, ultrafiltration, dialysis and so on [1-7]. The permeability is much dependent on both of chemical and physical structures of the membranes. The choice of the barrier materials for membranes and the control of their morphology are important to get effective permselective membranes.

Since Loeb and Sourirajan [8] found how to cast asymmetric cellulose acetate membranes, which consist of a very thin surface layer, supported by a more porous thick layer in 1962, many workers have investigated the preparation and performance of cellulose acetate membranes.

Polyamides and their analogue are also effective for the selective membranes and there have been developed many kinds of permselective membranes. In early 1960's, du Pont started to investigate the membranes for demineralization of water by reverse osmosis. After screening polymers, aromatic polyamides and polyhydrazides were shown to have superior properties [9-11]. In the present review various polyamides and their analogue are in focus as barrier materials for membranes, and their permeative characteristics will be discussed from the view point of their chemical structures.

Table 1. Correlation of Calculated (H_2O_{cal}) and Experimental (H_2O_{exp}) Water Sorption of Polyamic and Polyimide(cycl

Calculation of water sorption [b]					
Repeating unit [a]	(3 H$_2$O)	(1.8 H$_2$O)	(1.5 H$_2$O)	(1.5 H$_2$O)	(0.8 H$_2$O)
P/3,5-DAB		1	2	2	
P/3,5-DAB (cycl.)		1			
P/p-ABH	1		2'	1	
P/p-PDA			2	2	
P/p-PDA (cycl.)					
P/2,4-TDB			2	2	1
B/4-MPD			2	2	1
P/4,4'-DDE			2	2	
P/4,4'-DDM			2	2	
B/4,4'-DDE			2	2	1
P/4,4'-DDM			2	2	

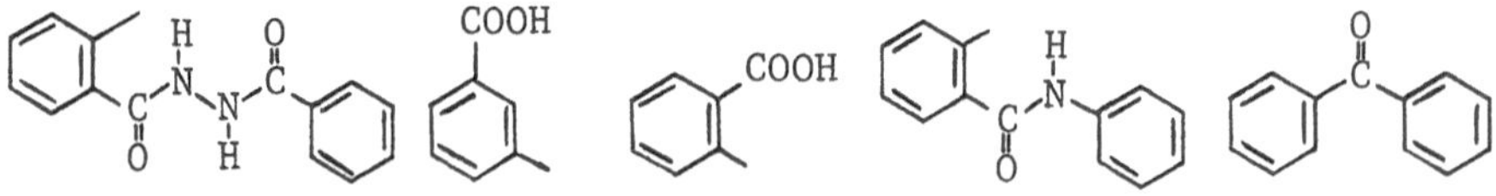

[a] P, pyromellitic dianhydride; B, 3,3',4,4'-benzophenone tetracarboxylic dianhydride; 3,5-DAB, 3,5-diaminobenz benzophenone; 4-MPD, 4-methoxy-m-phenylene diamine; 4,4'-DDE, 4,4'-diaminodiphenyl ether; 4,4'-DDM, 4
[b] The figure in parenthesis is the number of water molecules estimated to be associated with the corresponding polar gro

2 Some Aspects of Barrier Membranes

The properties desirable for permselective membranes are listed as follows:
1) Capability of forming thin barriers with high mechanical strength
2) High water permeability
3) High solute rejection (salt rejection)
4) Resistance to chemical and biological attacks
5) Low cost.

Roughly speaking, polyamides and their analogue appearing in the present review, as well as cellulose acetate, have substantially satisfied these conditions, but there will be still room for improvement.

High permeability of the membranes may essentially be realized mainly by 1) increase of the porosity, 2) decrease of the thickness, 3) increase of the operating pressure, and 4) increase of the hydrophilicity. The former three are on the processing and the latter two contain the subject of the properties of barrier materials.

The mere preparation of porous membranes is accompanied with a noticeable decrease of permselectivity [11], which is undesirable for reverse osmosis and ultra-filtration. A thin dense layer should be adopted to attain a high permeability with — out the decrease of permselectivity, but this necessarily decreases the mechanical strength. This conflict is largely resolved by the construction of asymmetric or composite membranes as described also in the present review.

embranes [13]

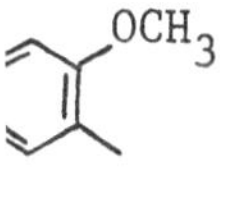	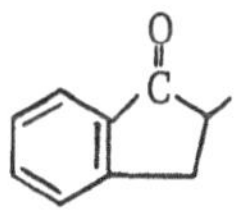	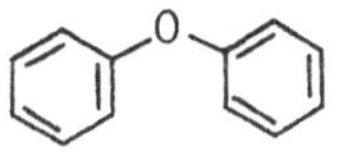	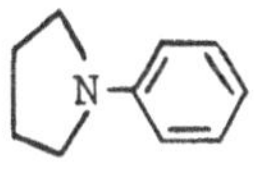	Total number of water molecules	H_2O_{cal}	H_2O_{exp}
OCH$_3$						
(0.8 H$_2$O)	(0.6 H$_2$O)	(0.5 H$_2$O)	(0.3 H$_2$O)			
				7.8	37.9	34.8
	4		2	4.8	25.9	28.4
				7.5	36.6	32.9
				6	33.1	30.9
	4		2	3	18.6	18.8
				10	32.7	29.3
				7.6	29.7	27.7
		1		6.5	28.0	21.5
				6	26.0	21.0
			1	7.3	25.2	19.1
				6	22.0	19.3

cid; p-ABH, p-aminobenzhydrazide; p-PDA, p-phenylene diamine; 2,4-TDB, 2,2',4,4'-tetramethoxy-5,5'-diamino-iaminodiphenylmethane;

Table 2. Average Number of Water Molecules in a Cluster[a] [20]

Polymer	Structural unit	Water sorption, g · H_2O/g · polymer	Water permeability, l/m² · day (kg/cm²) · μ	Salt rejection, %	Average number of water molecules in a cluster[b]: $1 + \varphi_1 G_{11}/\bar{v}_1$
Nomex		0.17	1.56	99.0	2.7
PA-6		0.21	1.72	99.8	2.9
PI		0.085	0.26	99.5	1.4
PB		0.22	7.4	99.8	2.7

PS	O=S=O with two benzene rings (diphenyl sulfone)	0.02	0.0086	20	5.0
Hydrin 100	$-CH-CH_2-O-$ $\quad\ \|$ $\quad CH_2Cl$	0.06	0.0086	—	8.2
Hydrin 200	$-CH-CH_2-O-CH_2-CH_2-O-$ $\quad\ \|$ $\quad CH_2Cl$	0.35	0.0086	—	8.8
Herchlor 21-79	$-CH-CH_2-O-CH_2-CH_2-O-$ $\quad\ \|$ $\quad CH_2Cl$	1.1	7.4	30	8.6

a Feed, 0.1 wt % NaCl aqueous solution;
b Zimm-Lundberg cluster function [21] was used for the estimation of these values

The higher hydrostatic pressures which prevail in reverse osmosis impose an additional mechanical requirement on the barrier materials, for example, an inherent polymer rigidity (stiffness) and a resistance to creep deformation. Aromatic compounds having high glass transition points should be more useful than aliphatic compounds for the membranes. Cross-linking is also an effective process to get networks of low chain mobility. Inter- and intramolecular interactions through polar groups such as hydrogen bonding should contribute to the mechanical strength of the membranes, which may be one of the reasons why polyamides are preferred as permselective membranes.

The hydraulic permeative coefficient of the membrane is generally expressed by the product of the diffusion constant and the solubility coefficient of water in the membrane. Hydrophilicity of the membrane increases the water solubility, resulting in its high water permeability [12]. Hydrophilic polyamides. therefore, can be regarded as effective materials for membranes. Their water flux can also be enhanced by the introduction of highly hydrophilic pendant $COOH$ or SO_3H groups. It is noteworthy that Walch et al. [13] were able to relate the solubility of water to the number of water molecules estimated to be associated with the polar groups and repeating units of polyamic acid and polyamides (Table 1).

The degree of crystallinity of the polyamides influences not only their processability but also their hydrophilicity. In high-crystalline polymers, the average interchain distance is very short, which principally results in the strong hydrogen bonding between amide groups. The $N—H\cdots O$ distance in α-helical polypeptides, aliphatic nylons, and all-para oriented aromatic polyamides is less than 3.1 A [14-17]. In these polymers, a casting solvent and water molecules hardly destroy the crystal structures formed by hydrogen bonding. In order to keep amide groups free, to interact with a polar solvent or water molecules, a design of some irregular structures as seen in copolymers and meta-oriented aromatic polyamides, or an introduction of bulky substituents is effective. In fact it is known that hydrogen bonds are not formed if the $NH\cdots O$ distance is greater than about 3.3 A [18]. The meta-phenylene isophthalamide with a $NH\cdots O$ distance of about 3.5 A [19] has a wide range of solubility. However, as hydrogen bonds contribute also to the mechanical rigidity and the wet strength of a membrane, it seems desirable to remain some degree of crystallinity.

The requirement of hydrophilicity in barrier materials has been widely accepted, but the mechanism by which it affects membrane performance, especially for the permselectivity, is not fully understood. Cellulose acetate and some kinds of polyamides and their analogues featured in the present review have both hydraulic permeability and permselectivity, while most highly hydrophilic materials have high permeability for water and show unselective permeation for ions and organic solutes.

Strathmann et al. [20] examined the water and salt sorption and the homogeneity of water distribution in various polymers and indicated that the uniformity of water distribution in the polymer is an important parameter controlling reverse osmosis desalination efficiency. As summarized in Table 2, the average number of water molecules included in a cluster is 1.4 to 2.9 for the superior barrier materials such as aromatic polyamides, polyamide-hydrazide, and polybenzimidazopyrrolone, while the number for the other polymer membranes is larger than 5.

Matsuura and Sourirajan [22-28] tried to examine various parameters of barrier materials such as the solute transport parameter ($D_{AM}/K\delta$), polar and nonpolar para-

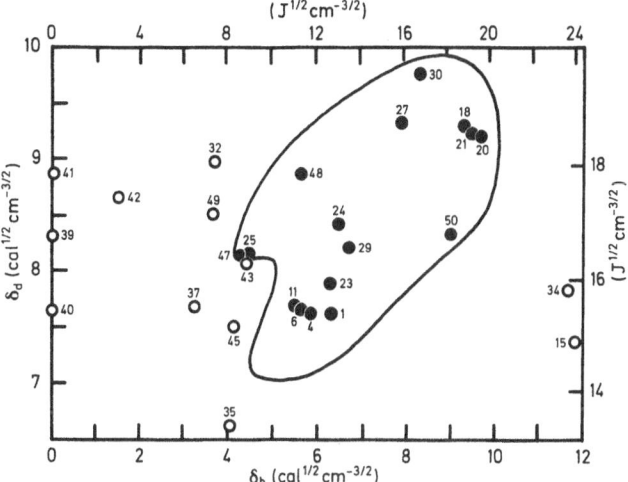

Fig 1. Relationship between δ_h and δ_d of various polymers [28, 29]

Table 3. Solubility Parameters of Various Polymers [29-31]

Polymer		δ_{sp}, cal$^{1/2}$ cm$^{-3/2}$	δ_d, cal$^{1/2}$ cm$^{-3/2}$	δ_h, cal$^{1/2}$ cm$^{-3/2}$
1	Cellulose acetate (CA 398)	12.65	7.60	6.33
4	Cellulose triacetate	12.03	7.61	5.81
6	Cellulose acetate propionate (CAP 151)	11.59	7.65	5.57
11	Cellulose acetate butyrate (CAB 171)	11.52	7.68	5.52
15	Cellulose	24.08	7.36	11.85
18	Aromatic polyamide	15.89	9.30	9.27
20	Aromatic polyhydrazide	16.25	9.20	9.60
21	Aromatic polyamide-hydrazide (PPPH 8273)	16.07	9.25	9.44
23	Polyurea (NS-100)	12.22	7.87	6.25
24	Sulfonated polyfuran (NS-200)	16.16	8.42	6.47
25	Polypiperazineamide (Pip-F)	12.07	8.15	4.50
27	Polybenzimidazolone (PBIL)	17.13	9.32	7.84
29	Nylon 6	12.43	8.45	6.66
30	Aromatic polyimide	19.00	9.75	8.23
32	Polysulfone	12.61	8.97	3.66
34	Polyvinyl alcohol	19.06	7.82	11.68
35	Polyvinyl formal	11.22	6.61	4.07
37	Polyvinyl butyral	9.83	7.67	3.25
39	Polyethylene	8.56	8.32	0
40	Polypropylene	8.02	7.65	0
41	Polystyrene	10.55	8.80	0
42	Polyvinyl chloride	11.03	8.65	1.45
43	Polymethyl methacrylate	9.93	8.08	4.40
45	Polyethylene glycol	9.37	7.50	4.14
47	Sulfonated polyphenyleneoxide	12.60	8.10	4.36
48	Sulfonated polysulfone	14.13	8.87	5.62
49	Polyacrylonitrile	14.39	8.51	3.65
50	Poly(ether/amide) (PA-300)	14.95	8.33	8.98

meters (α_p and α_n) and β-parameter estimated by using liquid chromatography, and proposed significant guidelines for the choice of membrane materials for reverse osmosis. They found the relationship between Hansen's solubility parameters of barrier polymers as shown in Fig. 1. [28,29] δ_h due to hydrogen bonding and δ_d due

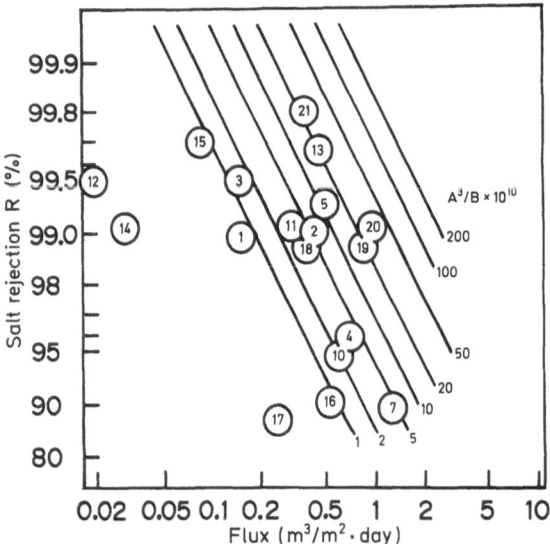

Fig. 2. Relationship between water flux and salt rejection of various polymer membranes [32]
A, flux × 3.32 × 10^{-5} (g/cm² · sec · atm);
B, A × (1 — R)/R × 34.8 (cm/sec);
condition, 40 kg/cm², 0.5 % NaCl aqueous solution, 25 °C.

1, Polyamide (Du Pont);	13, Poly N-amideimide (UOP);
2, DP-1 (Du Pont);	14, Polybenzooxadinon (Bayer);
3, Polyamide (Chemstrand);	15, Polyhidantoin (Bayer);
4, Polyamide (Monte Edison);	16, Sulfonated polysulfone;
5, Polyamide carboxylic acid (Toray);	17, Sulfonated PPO (GE);
7, DP-1 chelate (Du Pont);	18, NS-100 (North Star);
10, PBI (Cellanese);	19, PA-100 (UOP);
11, PBIL (Teijin);	20, PA-300 (UOP);
12, Polyamide (Forschung);	21, NS-200 (North Star)

to dispersion forces. The figures in Fig. 1 correspond with labels of the sample polymers in Table 3. The values of δ_h and δ_d can be estimated from the gross solubility parameters (δ_{sp}) and cohesive energies [29-31]. All polymers useful for barrier materials seem to be in the range of δ_h = 4 ~ 10 and δ_d = 7 ~ 10. The balance of hydrophilicity and hydrophobicity is important for control of the permselectivity. Kurihara [32] has summarized the water flux and salt rejection of various polymers as shown in Fig. 2, which is useful for the comparison of the permeative characteristics of barrier materials.

3 Aliphatic Polyamides

Polyamides, in general, are rather hydrophilic due to the interaction between polar amide groups and water molecules. Their mechanical strength is also enhanced by the molecular interactions through the hydrogen bonds between the amide groups.

Linear aliphatic polyamides came into use in du Pont for thin films and hollow fibers having high water permeability [33, 34]. Their permselectivity and mechanical strength were not sufficient for reverse osmosis, but have been developed as an ultrafiltration membrane with a high flux by making them asymmetric. Nylon 6 membranes prepared from formic acid solution containing $1 \sim 7\%$ polyethyleneglycol had an ultrafiltration capacity of 13,800 $1/m^2 \cdot hr \cdot bar$ and were 70 μ thick [35]. Nylon 4 membrane was recently shown no to exceed 53.3% of salt rejection on the reverse osmosis test using a 0.1% NaCl solution [36]. Uragami et al. [37] examined the permeation characteristics of nylon 12 membranes for ultrafiltration and reported that heat treatment of the membrane in aqueous solutions of mixed formic acid and formalin improved the permeability and the breaking strength.

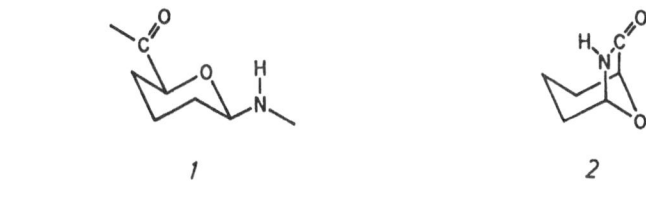

1   *2*

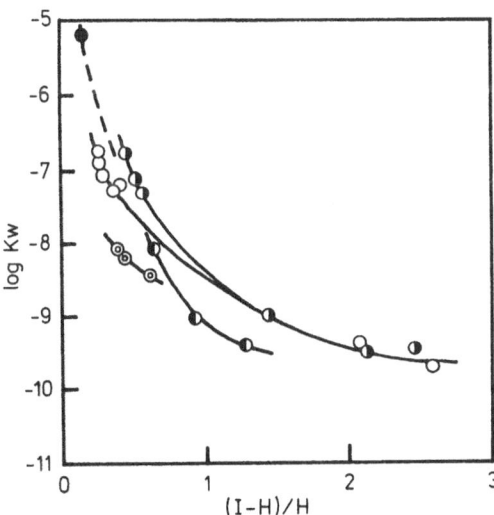

Fig. 3. Dependence of hydraulic permeability of various membranes on their degree of hydration H at 25 °C.

- ◐ polyBOL (casting from $CHCl_3$—CH_3OH solution);
- ○ polyBOL (casting from Me_2SO solution);
- ◑ optically active polyBOL (casting from CF_3CH_2OH + $CHCl_3$ solution);
- ● optically active polyBOL (casting polymerization in Me_2SO);
- ◎ crosslinked cellulose [43]

Table 4. Per Cent Solute Rejection from Its Aqueous Solution by Various PolyBOL Membranes[a 41)]

Solute	Mol. wt.	Solute rejection, %			
		Optically active		Racemic	
		M-5-2[b]	M-5-1[c]	R-3[d]	R-6[e]
Creatinine	113	3	2	3	28
Vitamin B$_{12}$	1,355	16	85	78	42
Albumin[f]	67,000	96	93	94	94

[a] Pressure, 3 kg/cm^2; temp., 25 °C;
[b] Optically active polyBOL membrane (immersed in water after casting); degree of hydration, 0.61; thickness, 0.10 mm; permeation rate, 1.6 l/m^2 · hr;
[c] Optically active polyBOL membrane (dried after casting); degree of hydration, 0.44; thickness, 0.040 mm; permeation rate, 0.18 l/m^2 · hr;
[d] PolyBOL membrane (casting polymerization); degree of hydration, 0.73; thickness, 0.58 mm; permeation rate, 1.9 l/m^2 · hr;
[e] PolyBOL membrane (dried after casting); degree of hydration, 0.27; thickness, 0.035 mm; permeation rate, 0.11 l/m^2 · hr;
[f] Bovine serum

A new class of hydrophilic polyamides, poly(tetrahydropyran-2,6-diylimino-carbonyl) *1* was prepared by the anionic polymerization of a bicyclic oxalactam (abbreviated as BOL, *2*) [38–42]. The resulting polyamide *1* has glass transition, fusion, and decomposition points at 130, 260–285, and 315 °C, respectively, and its membrane can be obtained by casting from a polyBOL solution. The solubility parameter

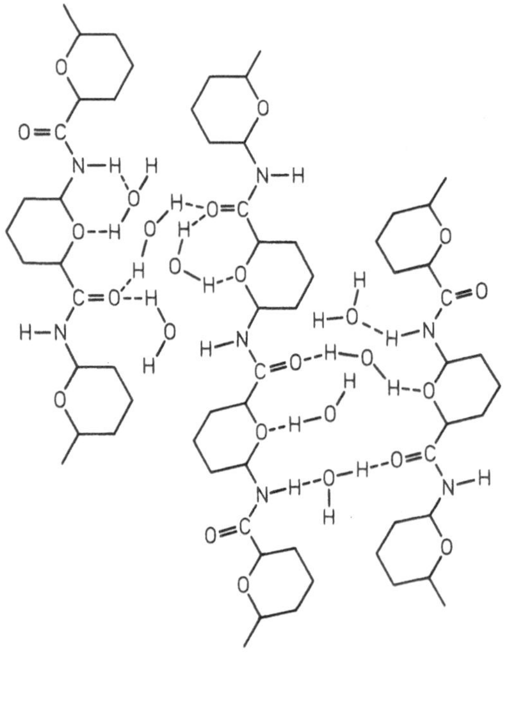

Fig. 4. Schematic representation of possible absorption modes of water molecules in hydrophilic microdomains in polyBOL membrane [42)]

of this polyamide is calculated to be 13.8 which is higher than that of cellulose acetate (Table 3). The hygroscopic polyBOL membrane exhibited an excellent behavior of fractional solute rejection in its aqueous solution, together with a high water permeability (Fig. 3 and Table 4).

Such permeation characteristics may be attributed to the novel polymer structure having an alternating and mosaic-like arrangement of hydrophilic and hydrophobic microdomains (Fig. 4). The hydrogen bonding between the amide groups should be loosened by the steric hindrance of bulky tetrahydropyran rings. There may also exist a large number of relatively weak polar site along the polymer chains. As a matter of course a large number of water molecules is easily absorbed. In fact the amount of bound water in the polyBOL membrane determined by differential scanning calorimetry was about three times more than that of nylon 6. Some of the possible absorption modes of water molecules at loosely-interacting polar sites such as amide and ether groups in the hydrophilic microdomains is schematically represented in Fig. 4 [42].

4 Polypeptides

Polypeptides obtained by the anionic polymerization of optically active N-carboxy-α-amino acid anhydrides are apt to have such an ordered structures as α-helices, which is useful for investigation on the relationship between the physical structure and the permeability of the membrane. Takizawa et al. [44-46] studied the water permeation and solute separation through poly(n-alkyl L-glutamate) membranes 3. It was concluded that water molecules permeate through relatively large free spaces

$$
\begin{array}{c}
\quad\quad\quad\quad O \\
\quad\quad\quad\quad \| \\
—— NH–CH–C —— \\
\quad\quad\quad\quad | \\
\quad\quad\quad\quad CH_2 \\
\quad\quad\quad\quad | \\
\quad\quad\quad\quad CH_2 \\
\quad\quad\quad\quad | \\
\quad\quad\quad\quad C{=}O \\
\quad\quad\quad\quad | \\
\quad\quad\quad\quad O \\
\quad\quad\quad\quad | \\
\quad\quad\quad\quad R \\
\quad\quad\quad\quad 3
\end{array}
\qquad
\begin{array}{l}
R = -CH_3, \ -CH_2CH_2CH_3, \\
\quad\ \ -CH_2CH_2CH_2CH_3
\end{array}
$$

between the helices. The water permeability for the polypeptide membranes was by about one order higher than that of cellulose acetate membranes and increased with the length of the side chain. The behavior was analyzed by the free-volume concept for mass transport. It was shown that the mechanism of water permeation involves the molecular transport of clusters consisting of several water molecules, though the size of the cluster is very small compared with that in bulk viscous flow. Although the rejection of sodium chloride was not so high as in cellulose acetate membranes, serum albumin was almost completely rejected in the aqueous solution.

5 Aromatic Polyamides and Polyhydrazides

The patents [9, 47)] filed by Richter and Hoehn in du Pont include many aromatic and heterocyclic condensed polymers for asymmetric reverse osmosis membranes, for example, the condensed polymers *4* from 3-aminobenzhydrazide, 4-amino-benzhydrazide (*5*), isophthaloyl chloride, and terephthaloyl chloride (*6*). The membranes were cast from polyamide solutions containing $0 \sim 3\%$ LiCl in dimethylacetamide,

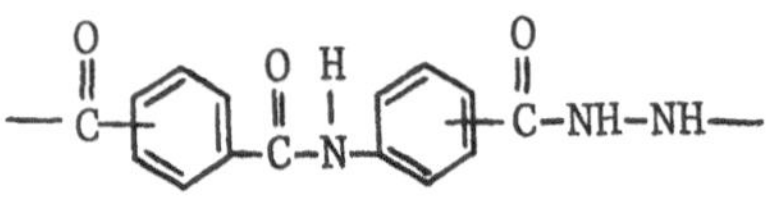

4 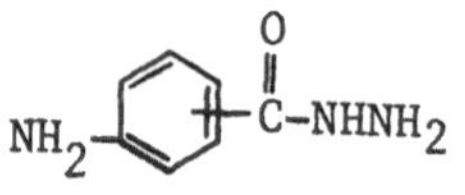   5 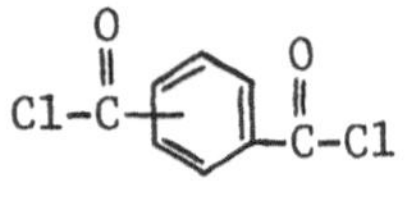   6

dimethylsulfoxide, and N-methyl-pyrrolidone. Applegate and Antonson [48)] obtained asymmetric polyamidehydrazide membrane (DP-1) from a solution of polymer *4* containing 3% LiNO$_3$, which showed a high water flux and a salt rejection ability similar to that of cellulose acetate membranes.

McKinney et al. [49, 50)] described the reverse osmosis properties of asymmetric membranes prepared from the following polyamides and polyamide-hydrazides (*7, 8,* or *9*).

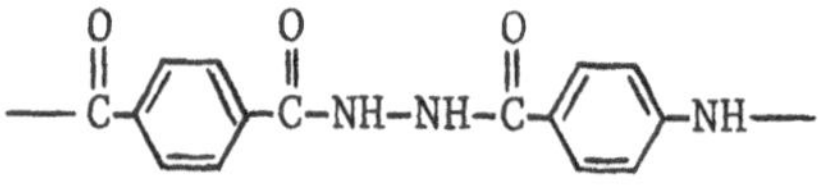

7

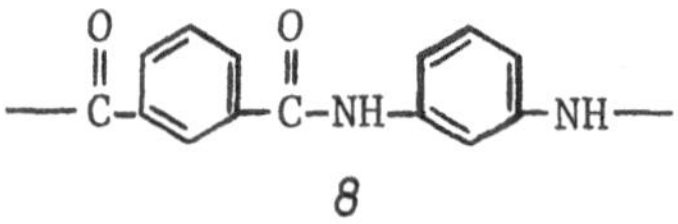

8

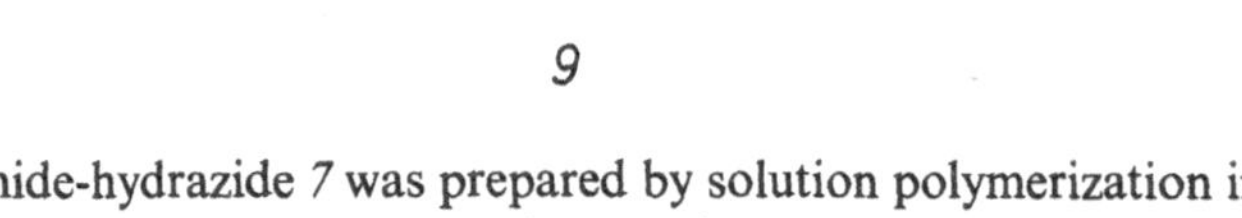

9

The polyamide-hydrazide *7* was prepared by solution polymerization in anhydrous dimethylacetamide from terephthaloyl chloride and p-amino-benzhydrazide at ca. 10 °C. The polyamide *8* resulted from the polycondensation of m-phenylenediamine with isophthaloyl chloride at —20 °C, whereas *9* was prepared by the reaction of terephthaloyl chloride with the complex diamine 1,3-bis(3-aminobenzamide)benzene at —20 °C. The water flux and salt rejection through these membranes were summarized in Table 5. The polyamide-hydrazide (*7*) membranes were prepared from polymer solutions containing $6 \sim 7\%$ polymer ($M_v \cong 34,000$) by casting on glass plates. The material was placed in an oven for $30 \sim 60$ min and coagulated in deionized water. Annealing of the solvent-free membranes at $40 \sim 100\ °C$ increased the membrane selectivity to sodium chloride. The polyamide membranes 8 and 9 were prepared in a manner similar to that used for the polymer 7 system, except that the annealing process was omitted for 9. A high degree of salt selectivity could be achieved as a direct result of the oven-controlled evaporative stage and subsequent coagulation.

Table 5. Reverse Osmosis Data for Aromatic Polyamides and Polyamide-hydrazides[a] [50]

Polymer	NaCl feed, %	Pressure, kg/cm²	Flux, l/m² day	Rejection, %
7	Sea water	105	285	95.7
8	0.5	70	489	99.2
8	Sea water	105	399	99.7
9	0.5	70	102	98.4

[a] AT 42 kg/cm²

Zschocke and Strathmann [51] obtained asymmetric reverse osmosis membranes from aromatic polyamides and polyamide-hydrazides and investigated their permeative characteristics. The polyamide membrane 10 cast from the dimethylacetamide containing pyridinium chloride demonstrated $99.0 \sim 99.5\%$ NaCl rejection with a water flux of $319 \sim 380\ l/m^2 \cdot day$ under pressure filtration at $100\ kg/cm^2$, while polyamide-hydrazide membrane 11 from the $LiNO_3$-containing dimethylacetamide solution exhibited quantitative rejection 99.8% for $MgSO_4$ but only $90 \sim 98\%$ rejection for NaCl with the water flux $200\ l/m^2 \cdot day$ under the same condition. The effect of the modification with pending carboxyl groups to the aromatic rings will be described later.

Low molecular weight salts contained in the casting solution of aromatic polyamides much increased the membrane fluxes without a detrimental effect on rejection, as shown in Table 6 [52].

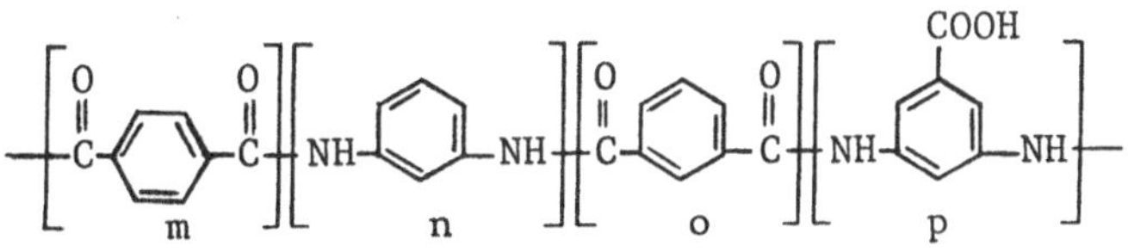

10
m/o=50/50, n/p=90/10

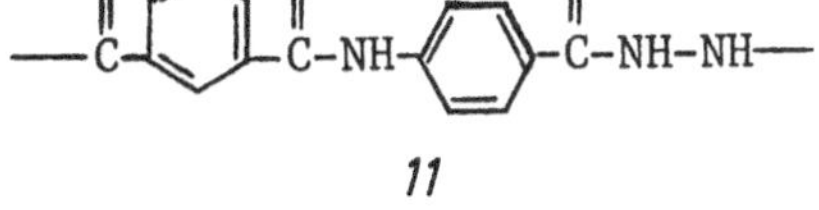

11

Highly dissociated salts, e.g. $LiClO_4$ or $Mg(ClO_4)_2$ exerted a stronger influence than the commonly used LiCl. With mixtures of different salts, stronger effects may be obtained than with a single additive. Many experimental facts indicated that the "salt effect" in aromatic polyamide membranes was due to the change in solvent activity, which had a large effect on the kinetics and equilibrium of solvent transport during evaporation and coagulation processes.

Table 6. Effects of Salts Contained in the Casting Solution on the Performance of Aromatic Membranes[a][52]

Salt	Salt concn., %	Rejection, %		Membrane flux, l/m² · day
		Urea	NaCl	
None	10	80	—	0.84
Pyridinium chloride	10	80	—	12.6
Pyridinium chloride	20	85	—	59
LiCl	5	77	—	34
LiClO₄	5	85	—	42
LiClO₄	10	78	—	51
ZnCl₂	10	83	99.0	46
Mg(ClO₄)₂	10	72	96.0	63
{ ZnCl₂ { Pyridinium chloride	{ 10 { 20	88	99.0	92
{ Mg(ClO₄)₂ { Pyridinium chloride	{ 10 { 20	80	99.9	189

[a] The polyamide used was obtained from a 1:1 mixture of m- and p-phenylenediamine and isophthaloyl chloride. All members were prepared under identical conditions (0.1 % aqueous solution, 50 kg/cm²)

Dickson et al. [53] suggested from the view points of the free-energy parameters for various alkali cations and halogen anions that the polyamide membrane surface behaved as if it was positively charged. Kinjo [54] in Hitachi Ltd. studied the ionic mobility in the membrane phase by membrane potential measurements. The asymmetric polyamide membranes were negatively charged, while the dense one was uncharged. He elucidated that the carbonyl groups of the polyamide chains in the skin layer were oriented to LiCl molecules in the casting solution and fixed in the skin layer during gelation.

The electron microscope has been used for the study of the structures of asymmetric polyamide membranes [49-52]. Jiayan et al. [55, 56] studied the morphology of aromatic polyamide type reverse osmosis membranes in detail by using scanning and transmission electron microscopes (SEM and TEM). The dense layer of the asymmetric membranes was prepared by a partial evaporation of solvent and immersion in gelating agent of a micellar structure, but various types of defects were observed on the surface of some parts of such membranes. The variation in structure in the cross-section for different membrane materials and preparation techniques was observed. The reverse osmosis performance correlated not only with the degree of densification before immersion for gelation during the preparation of the membrane, and the absence of pores in the surface layer of the resulting membrane, but also with the structure of the cross-section.

6 Aromatic-Aliphatic Polyamides

The polyamide membranes prepared from aromatic diamines and aliphatic dichlorides, or from aliphatic diamines and aromatic dichlorides may be feasible for ultrafiltration as well as aliphatic polyamide membranes. Ohya [57] investigated the separation characteristics of asymmetric poly(xylyleneadipamide) (12) membrane under severe conditions of high temperature and high (or low) pH.

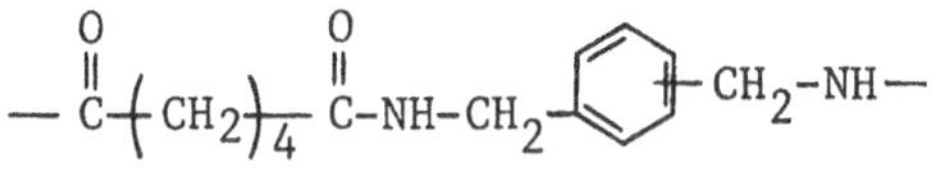

$$-\overset{O}{\overset{||}{C}}(CH_2)_4\overset{O}{\overset{||}{C}}-NH-CH_2-\!\!\!\!\!\!\!\!\!\bigcirc\!\!\!\!\!\!\!\!\!-CH_2-NH- \qquad para/meta=70/30$$

12

Walch et al. [58] in Hoechst fabricated a membrane for hemofiltration from 2,2,4-trimethylhexamethylenediamine-2,4,4-trimethylhexamethylenediamine-terephthalic acid copolymer *13*. The membrane has an ultrafiltration capacity for cattle blood of 1640 l/m² · day · bar, a retention value of 44 % for dextran 7000, and molecular weight limit of 58000 ± 6000.

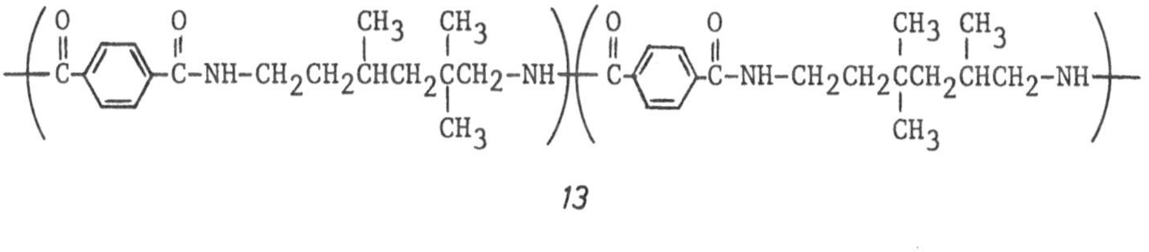

13

Vogl et al. [59, 60] prepared regular copolyamide membranes. They condensation-polymerized isophthaloylchloride with N,N′-bis(2-aminoethyl)oxamide, which was prepared from ethylenediamine and diethyloxalate, to give the polymer *14*. The salt rejection of the membrane was 77.5 ~ 99.0 % for 1 % aqueous solution of NaCl at 25 °C and 70 kg/cm² but the water flux was not very high.

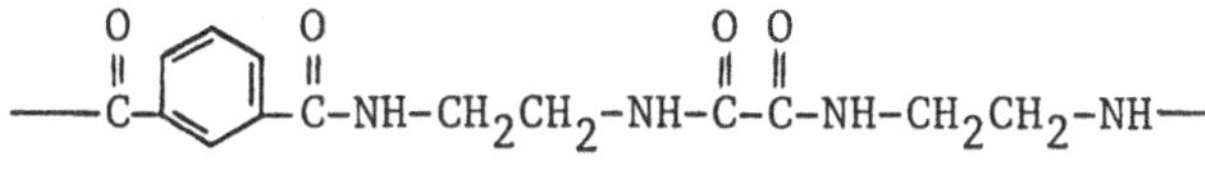

14

The chemical sensitivity or life expectancy of reverse osmosis membranes is very important for manufacturing application. Thus chlorine is the most well known reagent for water disinfection. Glaster et al. [61] inspected the influence of halogens on the performance and durability of reverse osmosis membranes. Cellulose acetate was unresponsive to halogen agents but polyamide-type membranes deteriorated rapidly when exposed to halogens.

7 Piperazine Polyamides

Credali et al. [62-65] prepared polypiperazineamides (*15*, R=trans-1,2-ethylenediyl, 2-propen-1,2-diyl, tetramethylene, m-phenylene, p-phenylene, thiofurazanediyl (*16*); R_1, H or CH_3; R_2, H or CH_3) membranes from the corresponding acyl dihalides and

15   16

piperazines. The membranes showed high compaction under pressure, high water permeability and chemical stability, which were attributed to the absence of amide protons. The desalination by the trans-2,5-dimethylpiperazine-*16* copolymer membrane were 98% with the water flux of 600 $l/m^2 \cdot$ day for 1% NaCl solution at 80 kg/cm^2, but salt rejections of the other polypiperazineamide membranes were lower than that of the former. Salt rejection by the trans-2,5-dimethylpiperazine-fumaroyl chloride copolymer membrane did not change markedly (94.4 ~ 96.4%) with decreasing of the film thickness from 39 to 4.0 μ, but the water flux increased from 15.5 to 140.3 $l/m^2 \cdot$ day. Hashida et al. [66] showed that poly(trans-2,5-dimethyl-piperazine-terephthalamide) *17* could be used for ultrafiltration membranes.

17

The membrane cast from chloroform-formic acid mixtures had an anisotropic structure with a 0.9–1.2 μ active layer and a 40 μ porous support layer. At a water flux of 139 $l/m^2 \cdot$ day (kg/cm^2 at 20 °C), the membrane showed 99.4% rejection of cytochrome C and 72.7% of Vitamine B_{12}. At 3980 $l/m^2 \cdot$ day water flux level, the rejection for bovine serum hemoglobin (MW, 66000 ~ 68000), cytochrome C, and Vitamine B_{12} were, 95.6, 79.4, and 39.8%, respectively.

8 Polybenzimidazols

Polybenzimidazols developed for heat-resistant resins were also applied to reverse osmosis membranes, first by Cellanese Corporation [67, 68]. The asymmetric poly-2,2'-(m-phenylene)-5,5'-dibenzimidazole membrane *18* was prepared from 3,3'-diamino-benzidine (*19*) and diphenyl isophthalate (*20*) and had a high water flux permeability, but the ability of salt rejection was not so high, although the heat treatment at 180 °C for 10 min improved the desalination ability.

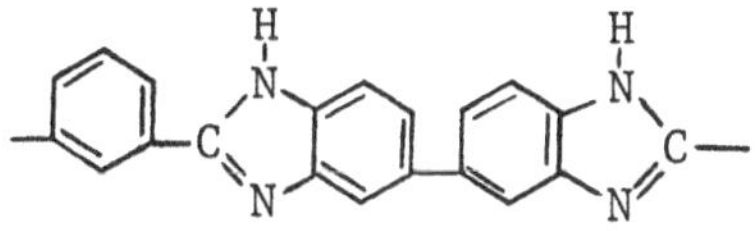

18

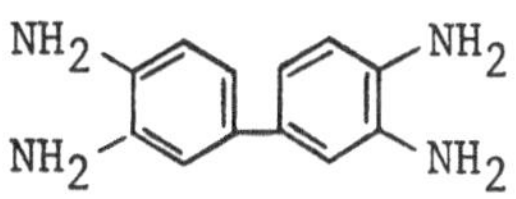

19

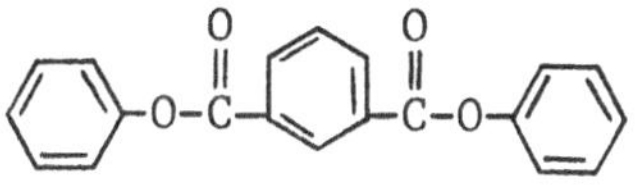

20

Polybenzimidazolone membrane *21* developed by Teijin Ltd. had the following permeative characteristics: Water permeation, 840 l/m² · day; salt rejection, 99.5% (1% NaCl aqueous solution, 80 kg/cm²) [69]. The membrane was less sensitive to plasticization with water than cellulose acetate and aromatic polyamide membranes

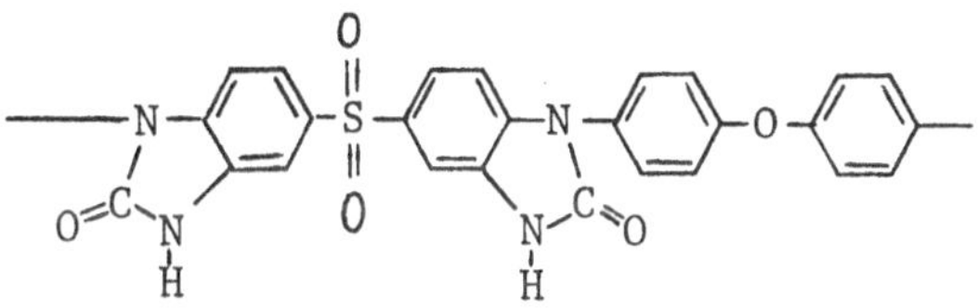

21

at high temperature, but less stable to impact. It was so highly resistant to chlorine that it remained stable even at pH 1.5.

9 Polysulfone-Polyamide

An oxidation-resistant polysulfone-polyamide membrane *22* was prepared by the reaction of equimolar amounts of 4,4′-diaminophenylsulfone and terephthaloyl chloride [70]. After soaking for 75 days at pH 1 ~ 2 in a 5 g/l CrO₃ solution, the membrane had a desalting ratio of >99% after a 260 hr continuous operation, while

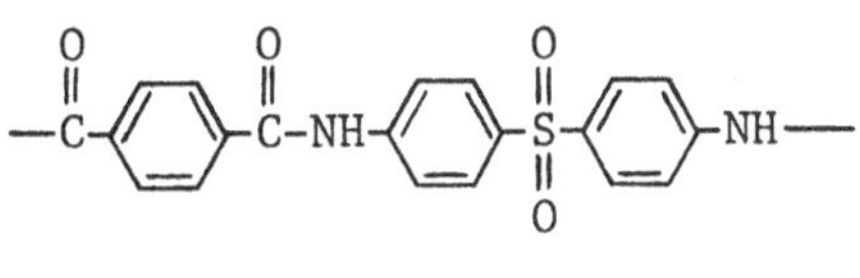

22

a polybenzimidazole membrane soaking in 0.2 g/l CrO₃ solution had 42–55% after 210 hr continuous operation.

10 Macrocyclic Polyether-Polyamides

Shchori et al. applied the polyether-polyamide [71] (PC-6, *23*) and its polymeric alloy containing 30% poly(vinyl pyrrolidone) to permeable membranes [72, 73]. NaSCN was strongly absorbed by PC-6. The permeative characteristics of PC-6 membranes were affected by their history because of reversible changes in the structure of the polymer network in the presence and absence of absorbed salts. Poly(vinyl pyrrolidone) had a stabilizing effect on the permeability of the membranes. The salt rejection values of the PC-6 membranes and membranes containing poly(vinyl pyrrolidone) were in the range of 95 ~ 99.5% and their permeabilities of water were at least one order of magnitude higher than those of the unmodified polyamides.

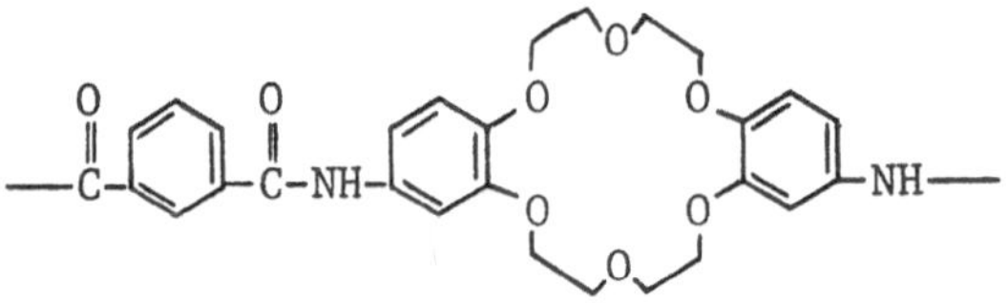

23

Frost [74] in Union Carbide Corporation also prepared polyether-polyamide membranes containing free carboxyl groups (*24*). The salt rejection of the membrane, however, was relatively low (46% for 3.5% NaCl aqueous solution at 70 kg/cm^2), although pure water flowed at 143 l/m^2 · day.

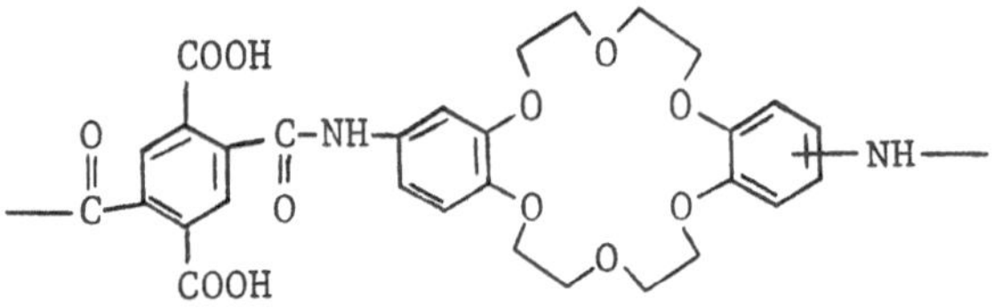

24

11 Phosphoramide-Carboxamide Copolymers

Kraus et al. [75–78] in Israel fabricated new aromatic phosphoric amide-carboxamide copolymers, which gave membranes high thermal stability, flame resistance, and salt rejection. Thus a copolymer was prepared from N,N'-bis(3-aminophenyl)-N''-phenylphosphoric triamide *25*, m-phenylenediamine and isophthaloyl chloride in

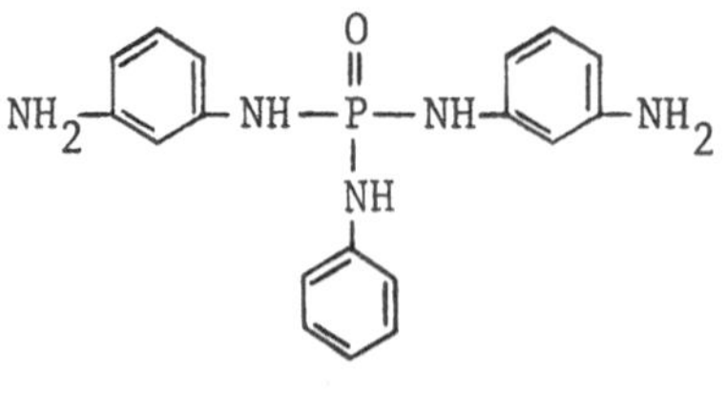

25

N,N-dimethylacetamide at -20 °C. A reverse osmosis membrane, formed by casting on a glass plate, had a salt and urea rejection of 96 and 94%, respectively, and a water throughput of 127 $l/m^2 \cdot$ day at pressure 50 kg/cm². The polymers of this family are distinguished by high fire resistance as demonstrated by high limiting oxygen indexes (LOI). The oxygen index, determined on a 0.1 mm membrane, was 46.9%. The polymers were stable for 2 hr at 400 °C. From preliminary tests the tensile strength of the new polymers was in the range of 800 $\sim$ 1100 kg/cm².

12 Polyimides

Wrasidlo [79] fabricated semipermeable poly(N-amide)imide (26) membrane from 3,3',4,4'-benzophenonetetracarboxylic dianhydride (27) and isophthalic dihydrazide (28). The membrane with 28 μ thickness had a water flux of 1140 $l/m^2 \cdot$ day-2000 A and salt rejection of 99.95%, while a 39.8% acetylated cellulose acetate membrane with 16 μ thickness had values of 118 $l/m^2 \cdot$ day and 99.35%.

Walch et al. [13, 80] determined the hyperfiltration properties of polyimide membranes 29. The polyamic acid-type prepolymer 30 were prepared by the addition

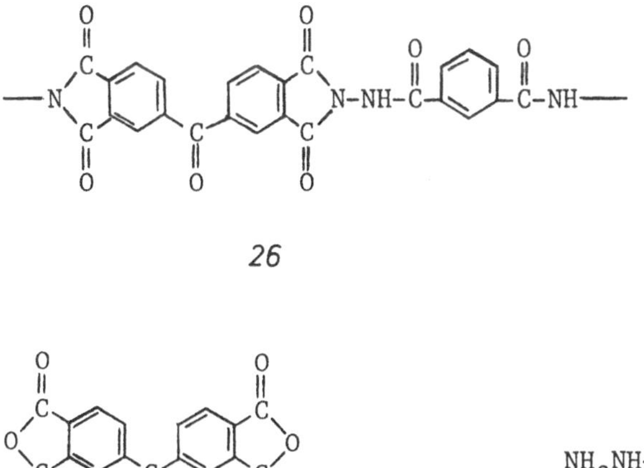

26

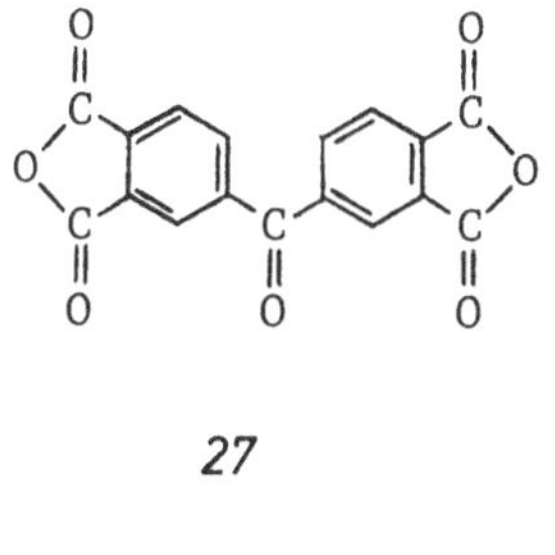

27

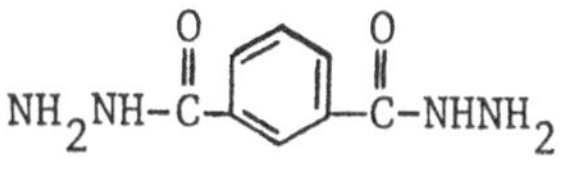

28

of a crystalline dianhydride to an equimolar solution of a diamine followed by the cyclization to the polyamides 29. They also found that the methoxylation of the aromatic polyamides increased the water flux without any significant loss of salt rejection. Methoxylation of polyamides improved their solubility in common organic solvents leading to a pronounced simplification of the processing of the membranes. Thus the water flux and NaCl rejection of pyromellitic dianhydride-m-phenylenediamine or pyromellitic dianhydride-4-methyoxy-m-phenylenediamine copolymers were 9 $l/m^2 \cdot hr$ and 99.9%, or 25 $l/m^2 \cdot day$ and 99.7%, respectively (membrane thickness, 12 μ; 100 kg/cm^2; 0.5% NaCl aqueous solution).

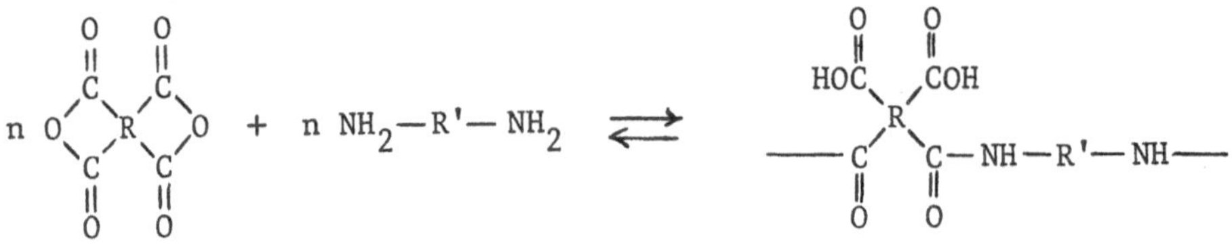

30

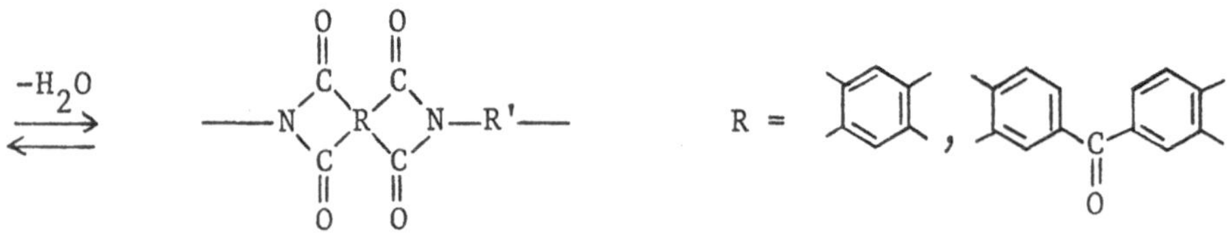

29

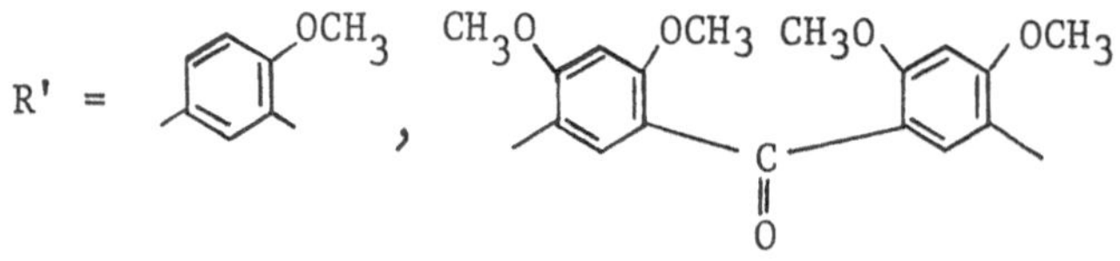

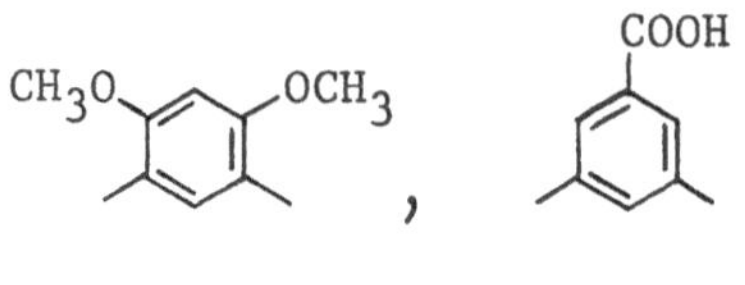

Recently Klimmek et al.[81] investigated new stable polyimide membranes for sea water desalination. Polyimide is generally stable to hydrolysis in the low pH range, but is apt to decompose through internal hydrolysis. They thought that the five-membered imide ring formed from pyromellitic anhydride and 4,4'-oxydianiline might have a steric tension, and fabricated the new polyimide membrane using 1,5-cyclooctadiene-1,2,5,6-tetracarboxylic acid anhydride and 4,4'-oxydianiline (Table 7). It is clear that the replacement of pyromellitic acid (6-membered ring) with cyclo-octadienetetracarboxylic acid (8-membered ring) leads to high resistance to hydrolysis of the resultant polyimide. Instead the water flux and the salt rejection decreased because the lipophilic cyclooctadiene-component favored an emergence of large water clusters. The use of the methoxylated component (4,6-dimethoxy-m-phenylene-diamine) for the membrane improved the water permeability and the ability of salt rejection.

Table 7. Permeation Characteristics of Some Polyimide Membranes [81]

Polymer structure	Water flux,[a] $l/m^2 \cdot day$	Salt rejection,[b] %	Stability to hydrolysis,[c] hr
(polymer structure diagram)	20	99.7	0.33
(polymer structure diagram with OCH$_3$, OCH$_3$)	310	99.8	0.13
(polymer structure diagram)	3.6	80.0	> 168
(polymer structure diagram with OCH$_3$, OCH$_3$)	69	93.4	5

[a] Reduced to 1 μm thickness;
[b] For 3.5% NaCl aqueous solution at 100 kg/cm^2 and 25 °C;
[c] In 0.1% KOH aqueous solution at 80 °C

13 Polyamides Having Pendant Polar Groups

Permeability of aromatic polyamide membranes have been improved by modification of aromatic rings with pendant polar groups, for examples sulfonic, carboxylic, carboxamide, and sulfonamide groups, in addition to the before-mentioned methoxy group.

Endoh et al. [82-89] in Toray Industry Inc. reported a study of new aromatic polyamides containing carboxylic groups and their fabrication into asymmetric membranes for reverse osmosis. The aromatic polyamides were prepared by low temperature solution polymerization from the combination of the following diamines and
acid derivatives: 4,4'-Oxydianiline, p(m)-phenylenediamine, 5,5'-methylenebis(2-
aminophenol), 3,5-diaminobenzoic acid, and iso(tere)-phthaloyl chloride, 4-chloro-
formylphthalic anhydride. [82] The films were cast from polymer solutions composed
of 10 ~ 20% polymer, 5 ~ 8% additives such as LiCl, LiNO$_3$, and solvents such as
N-methylpyrrolidone, dimethylformamide and dimethylacetamide. The use of LiNO$_3$
as additive in the casting solution generally produces membranes with a high salt
rejection and low water flux. LiCl alone can produce membranes with a higher flux
but with the salt rejection below the desired level. Consequently the combination
of the two salts was used in order to arrive at a suitable balance. The performance
characteristics of the high flux-type membrane 31 obtained from m-phenylenediamine and 4-chloroformylphthalic anhydride were 1330 l/m^2 · day and 97% salt
rejection, while the high rejection-type membrane 31 obtained from phenylenediamine
(para/meta = 50/50) and 4-chloroformylphthalic anhydride gave water flux 620 l/m^2
× day and salt rejection 99.3% under 40 kg/cm^2 to 0.5% NaCl aqueous solution
at 25 °C. They exhibited sufficient resistance in the tests carried out in the pH range
of 4 ~ 10.

31

Another copolyamide containing carboxyl groups was prepared from 3,3'-methy-
lenebis (anthranilic acid) 32, m-phenylenediamine, isophthaloyl chloride and terephthaloyl chloride [89]. The reverse osmosis membrane from this copolyamide was made
by treatment with transition metals (such as NiSO$_4$), Al, or Pb and shaping.

32

Reverse osmosis membranes were also prepared from polyamides with pendant
carboxamide groups [90]. For example, 4,4'-diaminodiphenylmethane-3,3'-dicarbox-
amide-isophthaloyl chloride copolymer 33 was dissolved in DMF containing LiCl,
cast to 250 μ thickness, dried at 100 °C for 15 min, and gelled in ice water to give
a membrane with the water flux permeability of 900 l/m^2 · day and salt rejection
of 80% (0.5% NaCl aqueous solution, 30 kg/cm^2). After heating the membrane in

water at 80 °C for 5 min, the water flux and salt rejection increased to 600 l/m² · day and 98 %, respectively.

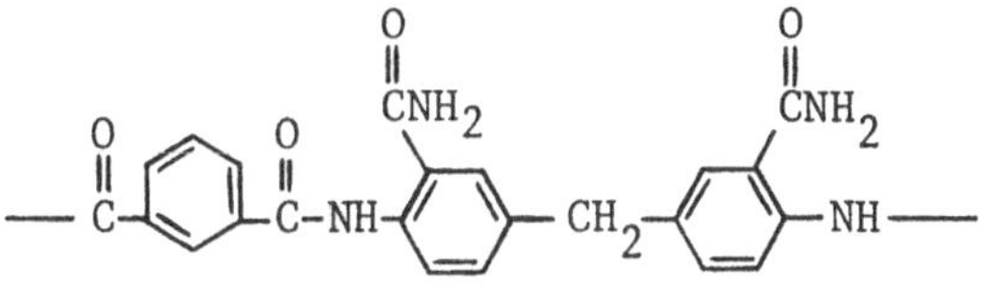

33

Heat-resistant polyamide membranes containing pendant sulfonamide groups were also fabricated [91−93]. Thus the membrane prepared from 2,2'-disulfonamide-4,4'-diaminodiphenyl ether-isophthaloyl chloride copolymer *34* gave the water permeation rate of 1700 l/m² · day and salt rejection of 65 %. The film with 50 μ

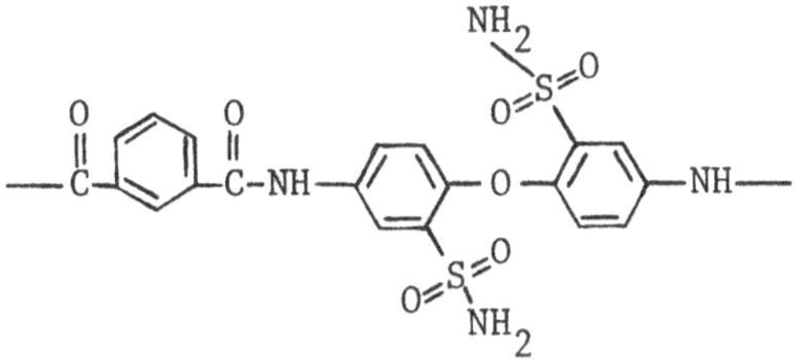

34

thickness had the tensile strength 1500 kg/cm² and was stable in air at <400 °C. A similar membrane containing N-methylsulfonamide groups [92, 93] had a high sugar rejection (90 %) with water flux of 430 l/m² · day.

14 Poly(amide imide) Having Pendant Sodium Sulfonate Groups

Sulfo group-containing poly(amide imide) semipermeable membrane *35* from trimellitic anhydride sodium 3,5-dicarboxybenzenesulfonate, and 4,4'-diphenyl-methane diisocyanate have been prepared [90]. A 110 μ-thick membrane gave water permeation 560 l/m² · day and salt rejection 99.0 % (0.5 % aqueous NaCl solution, 25 °C, 42 kg/cm²), while a trimellitic anhydride-4,4'-diphenylmethanediisocyanate copolymer membrane gave 30 l/m² · day and 98.2 %, respectively.

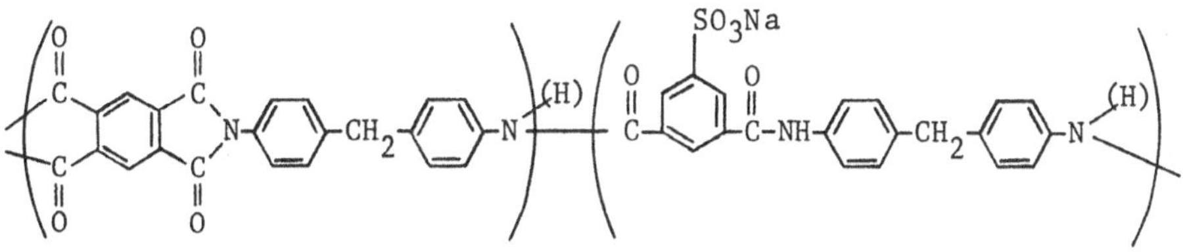

35

Kawahara et al. [95] fabricated an aromatic poly(amide imide sulfonamide) reverse osmosis membrane *36* from N-4'-chlorosulfonylphthalimide and p-phenylenediamine.

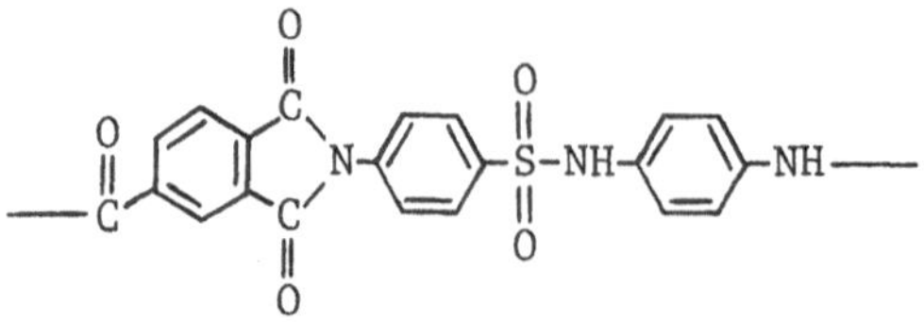

36

They prepared also a similar copolymer membrane having pendant sodium sulfonate groups *37* by using disodium 2,2'-benzidinedisulfonate in addition to the above-described materials [96]. The water permeability and salt rejection of the latter mem-

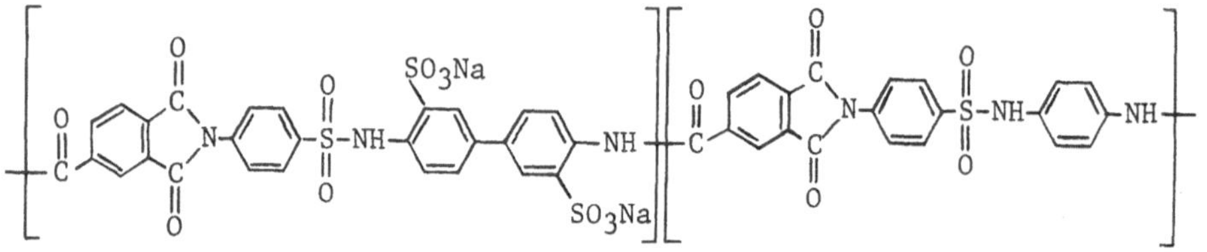

37

brane *37* were 530 l/m^2 · day and 98.6%, while those of the former membrane *36* were 90 l/m^2 · day and 98.0%, respectively.

As described above, the modification of the polyamide membrane with pendant polar groups effectively improved their permeation characteristics.

15 Thin-Film Composite Membranes

The ultrathin dense barrier layer and the thick porous underlayer can be separately fabricated from various materials and laminated together to give an asymmetric membrane. Thus a thin barrier membrane can be formed on the porous matrix by casting from a polymer solution, in-situ polymerization, or in-situ-interfacial condensation polymerization.

Caddote et al. [97–100] showed that polysulfone had a combination of compaction resistance and surface microporosity, and fabricated the following thin-film composite membranes by in-situ interfacial polymerization on the surface of the microporous support film. The NS-100 membrane was made by impregnating a polysulfone support with a 0.67% aqueous solution of polyethyleneimine, draining away excess reagent, then contacting the film with a 0.1% solution of tolylene-2,4-diisocyanate in hexane. An ultrathin polyurea barrier layer formed at the interface. This membrane was then heat-cured at 110 °C. A polyamide membrane (NS-101) was also fabricated by using isophthaloyl chloride in place of tolylene-2,4-diisocyanate. After impregnation of polyethyleneimine, the polysulfone support was dipped in 0.5% isophthaloyl chloride solution for 1 min. The water permeation of the resulting membrane was 1640 l/m² · day and salt rejection was 99.2%.

The poly(ether/amide) thin film composite membrane (PA-100) was developed by Riley et al., and is similar to the NS-101 membranes in structure and fabrication method [101, 102]. The membrane was prepared by depositing a thin layer of an aqueous solution of the adduct of polyepichlorohydrin with ethylenediamine, in place of an aqueous polyethyleneimine solution on the finely porous surface of a polysulfone support membrane and subsequently contacting the poly(ether/amide) layer with a water immiscible solution of isophthaloyl chloride. Water fluxes of 1400 ~ 1600 l/m² × day and salt rejection greater than 98% have been attained with a 0.5% sodium chloride feed at an applied pressure of 28 kg/cm². Limitations of this membrane include its poor chemical stability, temperature limitations, and associated flux decline due to compaction.

In pursuit of a chlorine-resistant, non-biodegradable thin-film-composite membrane, Cadotte et al. [97, 103, 104] fabricated interfacially the poly(piperazineamide) membrane (NS-300). The interfacially formed piperazine isophthalamide and terephthalamide membranes exhibited high salt rejection (98%) in sea water tests but their flux was low (Table 8). The replacing of the isophthaloyl chloride with its triacyl chloride analog, trimesoyl chloride improved vastly the flux of the membrane but its seawater salt rejection was low — in the range of 60 ~ 70% (*38*). The trimesoyl

38

Table 8. Performance of NS-300 Membranes in Reverse Osmosis Tests [97]

Acid chloride ratio[a]		Reverse osmosis test results			
		0.5% MgSO₄[b]		3.5% Synthetic seawater[c]	
Isophthaloyl : Trimesoyl		Flux l/m² · day	Salt rejection %	Flux l/m² · day	Salt rejection %
100	0	160	99.0	810	98
75	25	2360	99.6	2970	78
67	33	3140	99.9	3830	65
25	75	1260	99.3	3910	64
0	100	1060	99.3	3260	68

[a] Aqueous phase: 1% piperazine, 1% Na₃PO₄, 0.5% dodecyl sodium sulfate;
Hexane phase: 1% acyl chloride.
[b] At 14 kg/cm²;
[c] At 105 kg/cm²

chloride could be mixed with isophthaloyl chloride to produce copolyamide barrier layers. Maximum water permeability characteristics were found at an approximate copolymer ratio of 67% isophthalic and 33% trimesic groups. The differences in magnesium sulfate versus sodium chloride rejection appear to be due to the charged nature of the membrane barrier layer, since it is rich in carboxylate groups. Salt passage through the NS-300 membrane may be controlled by anions. In order to avoid the presence of free carboxylate groups in the membrane, piperazine-terminated oligomers, whose degree of polymerization was less than ten, were prepared from trimesoyl chloride and excess piperazine, and used in the interfacial reaction instead of piperazine. But fluxes of resulting membranes were low. In addition the continuous performance was accompanied by compaction, which decreased the permeability and permselectivity. In summary this membrane type may find applications in the desalination of blackish sulfate ground waters or industrial process waters, and may have utility in food application such as sucrose and lactose concentration.

Kawaguchi et al. [105] in Teijin Ltd. prepared a similar polyamide composite membrane from piperazine, trimesoyl chloride, and isophthaloyl chloride on a polysulfone support. The membrane exhibited high chlorine-resistance and excellent pressure-resistance. When used for reverse osmosis of an aqueous solution of 0.5% NaCl and NaOCl (available Cl 4 ~ 5 ppm) at pH 6.5 ~ 7.0, 25 °C, and 42,5 kg/cm^2, the water permeation was 1400 and 1330 l/m$^2 \cdot$ day and desalination was 93.4% and 95.7% after 2 and 100 hr, respectively.

They fabricated another two kinds of composite membranes through the interfacial reaction of triethylenetetramine [106, 107]. The one was the β,β'-dichloroethylether-triethylenetetramine-isophthaloyl chloride-trimesoyl chloride copolymer membrane, which had the water permeation rate of 2400 l/m$^2 \cdot$ day and desalination rate of 96.8%. The other was the adipic-triethylenetetramine-isophthaloyl chloride copolymer membrane, which showed the water flux 95.8 l/m$^2 \cdot$ day and NaCl rejection 99.8% on the reverse osmosis of a 0.5% aqueous solution at 25 °C and 42.5 kg/cm^2. These characteristics for both membranes did not decrease during the continuous operation for 100 ~ 500 hr.

Cadotte et al. [108] discovered a new thin-film composite membrane (FT-30), but its composition has not been disclosed.

16 References

1. Kesting, R. E.: Synthetic Polymer Membranes, McGraw-Hill, New York, N.Y. (1971)
2. Reverse Osmosis Membrane Research: Edited by Lonsdale, H. K. and Podall, H. E., Plenum Press, New York (1972)
3. Reverse Osmosis and Synthetic Membranes. Theory-Technology-Engineering: Edited by Sourirajan, S., National Research Council, Canada (1977)
4. Scott, J.: Membrane and Ultrafiltration Technology-Recent Advances, Noyes Data Corporation, New Jersey (1980)
5. Synthetic Membranes, Vol. 1, ACS Symp. Ser. *153* (1981)
6. Synthetic Membranes, Vol. 2, ACS Symp. Ser. *154* (1981)
7. Staude, E.: Angew. Makromol. Chem. *109/110*, 139 (1982)
8. Loeb, S. and Sourirajan, S.: Adv. Chem. Ser. *38*, 117 (1962)
9. Richter, W. J. K. and Hoehn, H. H.: Ger. Offen. *1*, 941, 932 (1970)
10. Beasley, J. K.: Desalination *22*, 181 (1977)

11. Lonsdale, H. K.: Desalination *13*, 317 (1973)
12. Yasuda, H., Lawase, D. E., and Peterlin, A.: J. Polym. Sci. A-2, *9*, 1117 (1971)
13. Walch, A., Lukas, H., Klimmek, A., and Pusch, W.: J. Polym. Sci., Polym. Lett. Ed. *12*, 697 (1974)
14. Holmes, D. R., Bunn, C. W., and Smith, D. J.: J. Polym. Sci. *17*, 159 (1955)
15. Inoue, K. and Hoshino, S.: J. Polym. Sci., Polym. Phys. Ed. *11*, 1077 (1973)
16. Northolt, M. G. and van Aartsen, J. J.: J. Polym. Sci., Polym. Lett. Ed. *11*, 333 (1973)
17. Northolt, M.: Eur. Polym. J. *10*, 799 (1974)
18. Hughes, E. W. and Moore, W. J.: J. Am. Chem. Soc. *71*, 2618 (1949)
19. Kakida, H., Chatani, Y., and Tadokoro, H.: J. Polym. Sci., Polym. Phys. Ed. *14*, 427 (1976)
20. Strathmann, H. and Michaels, A. S.: Desalination *21*, 195 (1977)
21. Zimm, B. H. and Lundberg, J. L.: J. Phys. Chem. *60*, 425 (1956)
22. Matsuura, T. and Sourirajan, S.: J. Coll. Int. Sci. *66*, 589 (1978)
23. Dickson, J. M., Matsuura, T., Blais, P., and Sourirajan, S.: J. Appl. Polym. Sci. *20*, 1491 (1976)
24. Matsuura, T., Brais, P., and Sourirajan, S.: J. Appl. Polym. Sci. *20*, 1515 (1976)
25. Matsuura, T., Brais, P., Paguan, L., and Sourirajan, S.: Ind. Eng. Chem., Process Des. Dev. *16*, 510 (1977)
26. Matsuura, T. and Sourirajan, S.: Ind. Eng. Chem., Process Des. Dev. *17*, 419 (1978)
27. Yeager, H. L., Matsuura, T., and Sourirajan, S.: Ind. Eng. Chem., Process Des. Dev. *20*, 451 (1981)
28. Matsuura, T. and Sourirajan, S.: Proc. Int. Symp. Fresh Water Sea 6th *3*, 227 (1978)
29. Matsuura, T.: Goseimaku-no-Kiso (Fundamentals of Synthetic Membranes), Kitami-Shobo, Bunkyo-ku, Tokyo (1981)
30. Hansen, C. M. and Beerbower, A.: Encyclopedia of Chemical Technology, Supplement Volume, Wiley, New York, 889 (1971)
31. Van Krevelen, D. W.: Properties of Polymers, Elsevier, Amsterdam (1976)
32. Kurihara, M.: Preprints of microsymposium on stiff chain polymers, Soc. Polym. Sci. Japan 63 (1978)
33. Lonsdale, H. K., Merten, U., and Riley, R. L.: J. Appl. Polym. Sci. *9*, 1341 (1965)
34. du Pont de Nemours, E. I.: Brit. Amended *1*, 177, 748 (1970)
35. Schindler, E. and Maier, F.: Ger. Offen. DE 3,028,213 (1982)
36. Huang, R. Y., Kim, U. Y., Dickson, J. M., Lloyd, D. R., and Reng, C. Y.: J. Appl. Polym. Sci. *26*, 1907 (1981)
37. Uragami, T., Maekawa, K., and Sugihara, M.: Desalination *27*, 9 (1978)
38. Sumitomo, H. and Hashimoto, K.: Macromolecules *10*, 1327 (1977)
39. Sumitomo, H., Hashimoto, K., and Ohyama, T.: Polym. Bull. *1*, 133 (1978)
40. Matsukura, T., Kinoshita, T., Takizawa, A., Tsujita, Y., Sumitomo, H., and Hashimoto, K.: Kobunshi Ronbunshu *35*, 803 (1978)
41. Hashimoto, K. and Sumitomo, H.: Macromolecules *13*, 786 (1980)
42. Sumitomo, H. and Hashimoto, K.: Contemporary Topics in Polymer Science, Vol. 4, Edited by Bailey, W. J. and Tsuruta, T.: Plenum, New York (1984)
43. Kawaguchi, M., Taniguchi, T., Tochigi, K., and Takizawa, A.: J. Appl. Polym. Sci. *19*, 2515 (1975)
44. Takizawa, A. and Hamada, T.: J. Appl. Polym. Sci. *18*, 1443 (1974)
45. Takizawa, A. and Hamada, T.: Polymer *15*, 157 (1974)
46. Takizawa, A., Taniguchi, T., Yamamuro, I., Tsujita, T., Nakagawa, T., and Minoura, N.: J. Macromol. Sci. — Phys. *B13*, 203 (1977)
47. Richter, J. W. and Hoehn, H. H.: Ger. Offen. 1,941,022 (1970)
48. Applegate, L. E. and Antonson, C. R.: Polymer Preprints *12* (2), 385 (1971)
49. McKinney, Jr., R.: Polymer Preprints *12* (2), 365 (1971)
50. McKinney, Jr., R.: Macromolecules *4*, 633 (1971)
51. Zschocke, P. and Strathmann, H.: Makromol. Chem. *73*, 1 (1978)
52. Kraus, M. A., Nemas, M., and Frommer, M. A.: J. Appl. Polym. Sci. *23*, 445 (1979)
53. Dickson, J. M., Matsuura, T., Blais, P., and Sourirajan, S.: J. Appl. Polym. Sci. *19*, 801 (1979)
54. Kinjo, N.: Maku *5*, 251 (1980)
55. Jiayan, C., Shuchun, B., Xingda, Z., and Lingying, Z.: Desalination *34*, 97 (1980)
56. Baoguan, Z., Guixiang, L., Jiayan, C., and Lingying, Z.: Maku *7*, 115 (1982)

57. Ohya, H., Negishi, Y., and Yamamoto, T.: Maku *3*, 143 (1978)
58. Walch, A., Wildhardt, J., and Beissel, D.: Eur. Pat. Appl. 2,053 (1979)
59. Vogl, O. F. and Stevenson, D. R.: U.S. Pat. Appl. 580,444 (1975)
60. Tirrell, D. and Vogl, O.: J. Polym. Sci., Polym. Chem. Ed. *15*, 1889 (1977)
61. Glaster, J., McCutchan, J. W., McCray, C. B., and Zachariah, M. R.: Synthetic Membrane, Vol. 1 (ACS Symp. Ser. *153*), 171 (1981)
62. Credali, L. and Parrini, P.: Polymer *12*, 717 (1971)
63. Credali, L., Parrini, P., Mortillaro, L., Russo, M., and Simonazzi, T.: Angew. Makromol. Chem. *19*, 15 (1971)
64. Credali, L., Chiolle, A., and Parrini, P.: Polymer *13*, 503 (1972)
65. Credali, L., Chiolle, A., and Parrini, P.: Desalination *14*, 137 (1974)
66. Hashida, I., Tanaka, M., and Nishimura, M.: Kobunshi Ronbunshu *34*, 763 (1977)
67. Model, F. and Lee, L.: ACS Preprints *32*, 384 (1972)
68. Model, F. S. and Lee, L. A.: Reverse Osmosis Membrane Research, Edited by Lonsdale, H. K. and Podall, H. E.: Plenum, New York, 285 (1972)
69. Hara, S., Mori, K., Taketani, Y., and Seno, M.: Proc. Int. Symp. Fresh Water Sea 5th *4*, 53 (1976)
70. Huan Ching Kó Hsueh: Peking University 1 (1978)
71. Feigenbaum, W. M. and Michel, R. H.: J. Polym. Sci., A-1 *9*, 817 (1971)
72. Shchori, E. and Jagur-Grodzinski, J.: J. Appl. Polym. Sci. *20*, 773 (1976)
73. Shchori, E. and Jagur-Grodzinski, J.: J. Appl. Polym. Sci. *20*, 1665 (1976)
74. Frost, L. W.: U.S. 3,956,136 (1976)
75. Kraus, M. and Nemas, M.: J. Polym. Sci., Polym. Lett. Ed. *19*, 617 (1981)
76. Frommer, M. A., Kraus, M. A., Nemas, M., and Gutman, R.: Ger. Offen. DE 3,018,826 (1982)
77. Frommer, M. A., Nemas, M., and Gutman, R.: Brit. UK Pat. Appl. GB 2,075,996 (1981)
78. Kraus, M. A., Frommer, M. A., Nemas, M., and Gutman, R.: U.S. 4,223,434 (1980)
79. Wrasidlo, W. J.: U.S. Pat. Appl. 316,936 (1972)
80. Walch, A., Klimmek, A., and Pusch, W.: J. Polym. Sci., Polym. Lett. Ed. *13*, 701 (1975)
81. Klimmek, A. and Krieger, W.: Angew. Makromol. Chem. *109/110*, 165 (1982)
82. Endoh, R., Tanaka, T., Kurihara, M., and Ikeda, K.: Desalination *21*, 35 (1977)
83. Endoh, R., Tanaka, T., Kurihara, M., and Ikeda, K.: Proc. Int. Symp. Fresh Water Sea 5th *4*, 31 (1976)
84. Ikeda, K., Susuki, T., and Bairinji, R.: Japan Kokai 73 22,358 (1973)
85. Ikeda, K. and Bairinji, R.: Japan Kokai 74 43,879 (1974)
86. Kurihara, M., Ikeda, K., Dokoshi, N., and Kobayashi, A.: Ger. Offen. 2,425,563 (1973)
87. Dokoshi, N., Kobayashi, A., Kurihara, M., Ikeda, K., and Itoga, M.: Japan Kokai 75 08,894 (1975)
88. Ikeda, K., Bairinji, R., and Dokoshi, N.: Ger. Offen. 2,308,197 (1974)
89. Tokizane, S., Kurihara, M., Tanaka, T., and Shimokawa, Y.: Japan Kokai 77 152,871 (1977)
90. Kinjo, N., Ishii, T., Miyadera, Y., and Yokono, H.: Japan Kokai 79 02,279 (1979)
91. Kinjo, N., Ishii, T., Yokono, H., and Nishidera, Y.: Ger. Offen. 2,729,847 (1978)
92. Kinjo, N., Ishii, T., Yokono, H., and Miyadera, Y.: Japan Kokai 79 89,982 (1979)
93. Kinjo, N., Ishii, T., Yokono, H., and Miyadera, Y.: Japan Kokai 79 86,596 (1979)
94. Nitto Electric Industrial Co. Ltd., Japan Kokai 81 24,007 (1981)
95. Kawahara, H. and Noda, A.: Japan Kokai 79 110,181 (1979)
96. Kawahara, H. and Noda, A.: Japan Kokai 79 110,182 (1979)
97. Cadotte, J. E. and Petersen, R. J.: Synthetic Membrane, Vol. 1, ACS Symp. Ser. *153*, 305 (1981)
98. Rozelle, L. T., Kopp, Jr., C. V., Cadotte, J. E., and Kobian, K. E.: Reverse Osmosis and Synthetic Membranes, Edited by Sourirajan, S.: National Research Council Canada, Ottawa, 249 (1977)
99. Cadotte, J. E.: U.S. 4,039,440 (1977)
100. Cadotte, J. E., King, R. S., Majerle, R., and Petersen, R. J.: Org. Coat. Plast. Chem. *42*, 13 (1980)
101. Riley, R. L., Fox, R. L., Lyons, C. R., Milstead, C. E., Seroy, M. W., and Tagami, M.: Desalination *19*, 113 (1976)
102. Riley, R. L., Milstead, C. E., Lloyd, A. L., Seroy, M. W., and Tagami, M.: Desalination *23*, 331 (1977)
103. Cadotte, J. E.: U.S. 4,259,183 (1981)

104. Cadotte, J. E.: Eur. Pat. Appl. 14,054 (1980)
105. Kawaguchi, T., Taketani, Y., Minematsu, H., Hayashi, Y., and Hara, S.: Japan Kokai 80 139,802 (1980)
106. Kawaguchi, T., Sasaki, N., Hayashi, Y., and Ono, T.: Japan Kokai 79 102,291 (1979)
107. Kawaguchi, T., Minematsu, H., Taketani, Y., Hayashi, Y., and Hara, S.: Japan Kokai 80 49,106 (1980)
108. Cadotte, J. E., Petersen, R. J., Larson, R. E., and Erickson, E. E.: Desalination *32*, 25 (1980)
109. Applegate, L. E. and Antonson, C. R.: Reverse Osmosis Membrane Research, Edited by H. K. Lonsdale and H. E. Podall, Plenum, New York, 243 (1972)

M. Gordon (Editor)
Received September 7, 1983

Membranes with Non-Homogeneous Sorption and Transport Properties

J. H. Petropoulos

Physical Chemistry Laboratory, Democritos Nuclear Research Centre, Aghia Paraskevi, Athens, Greece

The present review deals with selected developments in a variety of fields of membrane and polymer research organized around the central theme indicated by the title. The emphasis throughout is on fundamental principles and examples which best illustrate these principles, rather than on an encyclopedic review. The membrane systems considered are of topical interest, on one hand, and exemplify various degrees of inhomogeneity of sorption and transport properties, on the other hand. Inhomogeneities at the microscopic level can affect sorption and transport properties markedly by creating new sorption modes. This is well exemplified by dual mode gas sorption and diffusion in glassy polymers; which is being actively studied also from a dual point of view, namely the fundamental understanding of the glassy state, on one hand, and the prediction of gas separation properties, on the other hand. Polyamide-acid dye systems have been quoted as another physically very different example, which is important in textile dyeing, but could prove important also from the biomimetic point of view. Non-homogeneity at the microscopic level is exemplified by the gas permeability of composite membranes and the ionic sorption and transport properties of "homogeneous" ion exchange and other membranes. In the former case, the problem of interest is the relation between the properties of the composite and those of the component phases. In the latter case, there are no recognizable component phases and the effects of non-homogeneity have often been interpreted in other ways. Finally, membranes exhibiting macroscopically nonhomogeneous sorption and diffusion properties are considered. The major development here concerns the characterization of such properties by means of transient-state measurements.

1 Introduction

The membranes to be considered here may be divided into three classes according to the scale of inhomogeneity which is reflected in their sorption and transport properties:

(a) On a nearly molecular scale, the membrane may possess microregions differing substantially in some property from the rest of the polymer matrix. These can act as specific sorption sites for penetrant molecules, which are also sorbed in a non-specific manner or "dissolved" by the polymeric matrix (treated as essentially homogeneous).

(b) On a higher, but still microscopic, level of nonhomogeneity, there may be in fact (or assumed to exist in theory) more or less distinct microregions or "domains" sufficiently extensive to constitute individual thermodynamic phases characterized by their own sorption and diffusion coefficients.

(c) Finally, the membrane may exhibit spatial variation in sorption and diffusion properties on the macroscopic scale.

Sorption and diffusion of a given penetrant in a homogeneous membrane may, in general, be described by [1]

$$S = C/a \tag{1}$$

$$\frac{\partial C}{\partial t} = \frac{\partial}{\partial X}\left(D_T S \frac{\partial a}{\partial X}\right) \equiv \frac{\partial}{\partial X}\left(P \frac{\partial a}{\partial X}\right) \tag{2}$$

where X denotes position along the direction of diffusion, i.e. across the membrane (with $X = 0, 1$ at the membrane surfaces, 1 being the thickness of the membrane); t is the time; $C(X, t)$ is the concentration of penetrant in the membrane and $a(X, t)$ its activity; S, D_T and $P \equiv D_T S$ are the (so called thermodynamic [1]) sorption (or solubility or partition), diffusion and permeability coefficients respectively. The activity of the penetrant in the membrane is here defined as equal to that of the penetrant in the contiguous external phase (which may, in turn, be approximately equated with the gas concentration or pressure, the relative vapour pressure, or the solute concentration, if the external phase is a sufficiently dilute gas, vapour or solution respectively) [1]. S measures the partition of penetrant between the membrane and external phases, D_T describes its mobility in the membrane, whereas the flux of penetrant across the membrane is governed by P. For a homogeneous membrane S and D_T are constants (i.e. Henry's and Fick's laws respectively are obeyed), if the membrane-penetrant system is thermodynamically ideal. If it is not, S will be a function of C (or a). D_T may also be a function of C, if the properties of the polymer are appreciably modified by the presence of penetrant (e.g. by plasticization) [2,3]. Under any of the above circumstances, it is possible to transform Eq. (2) into the well known form

$$\frac{\partial C}{\partial t} = \frac{\partial}{\partial X}\left(D \frac{\partial C}{\partial X}\right) \tag{3}$$

which defines a "practical" (or "Fick") diffusion coefficient D related to D_T by

$$D = D_T S \frac{da}{dC} = \frac{D_T}{1 + d\ln S/d\ln a} \tag{4}$$

S and D (and hence D_T) or P for any particular membrane-penetrant system can be determined experimentally by standard methods based on Eq. (1) and appropriate full or partial solutions of Eq. (3) [2,3].

Membranes classified under (a) and (b) above may be fully characterized by overall (macroscopic) coefficients S and D_T or P for all experimental purposes exactly as above. The only difference from strictly homogeneous membranes is that the said macroscopic coefficients are of limited physical significance until they are analyzed into their component microscopic coefficients.

Treatment of class (c) membranes, on the other hand, presents a considerably more complicated problem. Here, S and D_T in Eqs. (1) and (2) are functions of the spatial coordinates. The problem becomes much more acute if S and D_T are also dependent on C [4,5]. Under these conditions, transformation of Eqs. (2) into (3) is not generally possible and there are no standard methods, as in the previous cases, of fully characterizing the membrane-penetrant system [3-5]. There is usually no difficulty in determining an overall or effective solubility coefficient; but the definition of useful effective diffusion coefficients is a more difficult matter, which, not surprisingly, is a major concern of current research in the field.

2 Membranes Exhibiting Multiple Sorption and Diffusion Modes

The sorption of a penetrant by a class (a) membrane can be treated most simply, if the processes of sorption in the polymer matrix and into each kind of specific sorption site within it can be treated as mutually independent and hence additive. Under these conditions, each sorbed species may similarly be expected to contribute additively to the total diffusion flux. Hence, if there are N −1 kinds of specific sorption sites, the overall sorption and permeability or diffusion coefficients can be written as

$$S = \sum_{i=1}^{N} S_i \tag{5}$$

$$P = \sum_{i=1}^{N} P_i = \sum_{i=1}^{N} S_i D_{Ti} \tag{6}$$

$$D_T = P/S = \sum_{i=1}^{N} S_i D_{Ti} \Big/ \sum_{i=1}^{N} S_i \tag{7}$$

Here, S_i and P_i or D_{Ti} characterize the ith sorbed penetrant species (i = 1 refers to the polymer matrix and i = 2, 3, ... , N identify each kind of specific sorption site).

Membrane-penetrant systems, whose sorption and diffusion properties can be described by Eqs. (5)–(7) with N = 2 ("dual mode sorption and diffusion models") have attracted much interest. The most important examples of such systems are considered in the next two sections.

3 Dual Mode Gas Sorption and Diffusion
in Glassy Polymer Membranes

3.1 Sorption of Single Gases

Earlier work on the application of the concept of dual mode sorption and diffusion to glassy polymer-gas systems has been reviewed in detail [6] and important aspects of more recent work have been dealt with in more recent reviews [7-10]. Eq. (5) was first applied by Michaels et al [11]. Sorption in the polymer matrix and in the specific sorption sites was represented by linear (Henry's law) and Langmuir isotherms respectively; so that S_i in Eq. (5) is given by

$$S_1 = K_1 \tag{8}$$

$$S_2 = s_0 K_2 / (1 + K_2 a) \tag{9}$$

where K_1, K_2 and s_0 are constants. In practice, a is usually replaced by the gas pressure p (and K_1, K_2, s_0 are usually replaced, in this field, by the symbols k_D, b and C_H' respectively).

Eq. (5) in conjunction with Eqs. (8) and (9) have, so far, provided adequate representation of experimental isotherms [6-32], which are characterized by an initial convex-upward portion but tend to become linear at high pressures. Values of K_1, K_2 and s_0 have been deduced by appropriate curve-fitting procedures for a wide variety of polymer-gas systems. Among the polymers involved in recent studies of this kind, one may cite polyethylene terephthalate (PET) [12-14], polycarbonate (PC) [19-22,27], a polyimide [16,17], polymethyl and polyethyl methacrylates (PMMA and PEMA) [18], polyacrylonitrile (PAN) [15], a copolyester [26], a polysulphone [23], polyphenylene oxide (PPO) [25], polystyrene (PS) [27,28], polyvinyl acetate [29] and chloride [32] (PVAc and PVC), ethyl cellulose [24] (EC) and cellulose acetate (CA) [30,31]. A considerable number of gases have been used as penetrants, notably He, Ar, N_2, CO_2, SO_2 and light hydrocarbons.

At the empirical level, alternative ways of representing the data (which may or may not utilize the dual mode concept) are possible [33,34]. Accordingly, interest in the model embodied in Eqs. (5), (8) and (9) centers primarily around the question of its physical meaningfulness and significance, as well as its predictive potential. The results of recent research will be discussed below mainly from this point of view.

A basic characteristic of glassy polymers is to be found in their "excess properties", such as excess enthalpy or volume. Thus, it is well known that densification of a polymer cooled significantly below its glass transition temperature T_g is a continuing, but very long term, process. Consequently, a glassy polymer ordinarily exists in a quasiequilibrium state characterized by specific volume $\hat{V}_g$ larger than the equilibrium value $\hat{V}_l$ (which corresponds to "complete densification"). It is often assumed that, at temperatures T not too far below T_g, $\hat{V}_l$ can be identified with the value obtained by extrapolation of the $\hat{V}(T)$ line measured above T_g. Hence, the fractional "excess free volume" is given by

$$v_E = (\hat{V}_g - \hat{V}_l)/\hat{V}_l = (\alpha_l - \alpha_g)(T_g - T) \tag{10}$$

where $\alpha_1, \alpha_g (\alpha_1 > \alpha_g)$ denote the thermal expansion coefficient measured above and below T_g respectively. This excess free volume is considered to be dispersed in the glassy polymer in the form of semipermanent, low density microregions of molecular dimensions ("frozen microvoids"). These act as specific sorption sites, because less energy need be absorbed to open up a hole large enough for accomodation of a penetrant molecule (and hence the sorption process is more exothermic) in these regions than in the (denser) polymer matrix [11]. It is then natural to associate the Henry sorption mode with the (more or less) fully densified matrix and to regard the site, or Langmuir-like, sorption mode as the "excess sorption" property of the polymer associated with v_E.

This picture fits very well with the tendency of the sorption isotherm curvature (and hence of the site sorption mode) to disappear at $T > T_g$. On a more quantitative level, the above characterization of the Henry sorption mode is supported by the smooth temperature dependence of K_1 found in the PET-CO_2 system [12], which indicates a roughly unchanged enthalpy of sorption $\overline{\Delta H}_1$ above and below T_g. Additional support is provided by the correlation between K_1 and the Lennard-Jones parameter ε/k characteristic of the gaseous penetrant, in accordance with

$$K_1/K_1^0 = \exp(m\varepsilon/k) \qquad (11)$$

where K_1^0, $m = $ const and $m \simeq 10^{-2}$ K^{-1} Eq. (11) holds, in fact for sorption in liquids and rubbers (cf. Table 1) [25].

The quantitative formulation of the site sorption mode, on the other hand, has the virtue of simplicity, but is undoubtedly rather highly idealised. Ideally, Eq. (9) refers to a collection of distinct, permanent and independent sites each accomodating one penetrant molecule; s_0 measures the concentration of these sites in the membrane and K_2 their affinity for the penetrant assuming them to be isoenergetic [35,36]. On this basis, the temperature dependence of K_2 should yield a constant enthalpy for this sorption mode $\overline{\Delta H}_2$ [35,36]. Consistency with the physical picture presented above requires moreover that $\overline{\Delta H}_2$ be more exothermic than $\overline{\Delta H}_1$ [11].

Table 1. Typical values of the parameters appearing in Eq. (11) for gases in a variety of amorphous media [25]

Material		$m \times 10^2$ ($^\circ K^{-1}$)	$K_1^0 \times 10^2$ (cm^3 (STP)/cm$^3 \cdot$ atm)	T ($^\circ$C)
Liquids:	Benzene	0.95	2.98	25
	n-heptane	0.94	1.69	25
Rubbers:	Natural rubber	0.94	1.11	25
	Silicone rubber	0.94	1.88	25
	Butyl rubber	1.00	0.90	25
	Amorphous polyethylene	0.94	0.72	25
	Polychloroprene	0.97	1.10	25
Glasses:	Polycarbonate	0.92	1.15	35
	Polysulfone	0.96	0.93	35
	Copolyester	0.96	1.03	35
	Poly(phenylene oxide)	0.93	1.65	35

In practice, it is not easy to judge how much reliance can be put on the enthalpies of sorption quoted (mostly for CO_2 in various polymers), which include the cumulative uncertainties of curve-fitting the relevant experimental isotherms and subsequent linearization of the corresponding Vant Hoff plots. With this reservation, we note that the condition $-\overline{\Delta H}_2 > -\overline{\Delta H}_1$ is fulfilled in some cases [6,12,18,22] but not in others [15,16,18,27]. The presence of solvent residues in the membranes used is thought to be largely responsible for the latter results [15,16,18]. Further clarification of this point is evidently necessary. Even so, there remains the more fundamental question as to the precise physical meaning of $\overline{\Delta H}_2$ values determined in the above manner. It is generally believed that the prerequisites for a true Langmuir sorption mechanism are unlikely to be fulfilled in real sorbents [27,33,35,36] and that formal conformity to a Langmuir isotherm may be observed even when the isoenergetic site assumption is demonstrably incorrect [35-37].

The parameter s_0 is also dependent on temperature [13,18,22,27,38]. Although this temperature dependence may be represented in terms of a (variable [14]) enthalpy [14,22], it is more natural to seek a correlation between s_0 (designated as a capacity parameter) and the excess free volume v_E. Impressive results have been obtained in this connection, on the basis of Eq. (10), mostly with CO_2 as penetrant. As illustrated in Fig. 1, s_0 tends to decrease nearly linearly with rising T and vanishes at $T > T_g$ [38]. Fig. 2 shows an analogous correlation between s_0 for CO_2 at $T = 35\,°C$ and T_g, using a wide variety of polymers in accord with Eq. (10) (since $\alpha_1 - \alpha_g$ varies relatively little) [25]. These findings are consistent with direct proportionality between s_0 and v_E. It is noteworthy, incidentally, that the dependence of s_0 on (i) polymer molecular weight

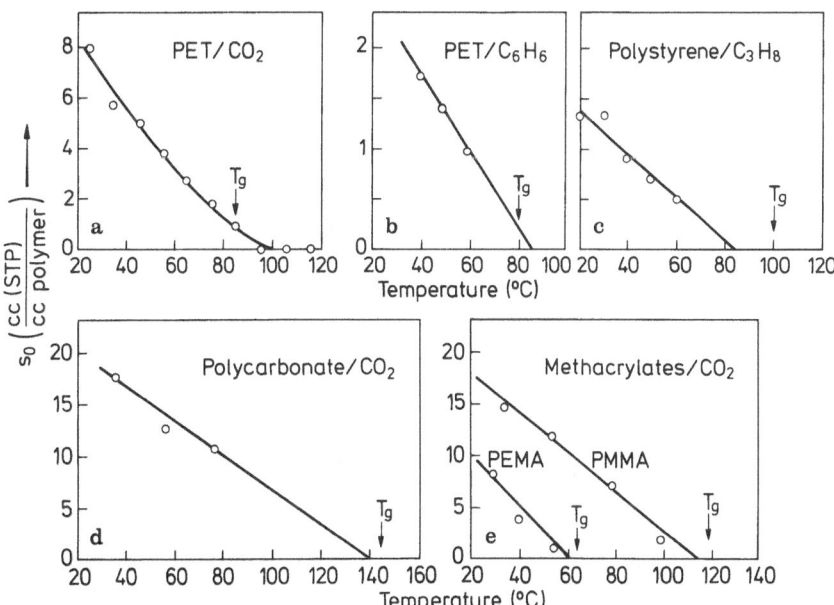

Fig. 1. Temperature dependence of the experimental values of the Langmuir capacity parameter s_0[cf. Eq. (10)] for [38] (a) PET-CO_2 (b) PET-C_6H_6 (c) PS-C_3H_8 (d) PC-CO_2 (e) PMMA, PEMA-CO_2

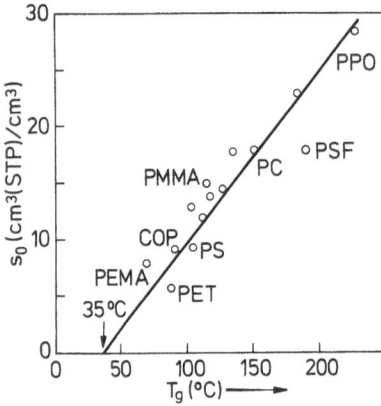

Fig. 2. Correlation of experimental values of the Langmuir capacity parameter (s_0) for CO_2 at 35 °C with the glass transition temperature of the polymer [25] [cf. Eq. (10)]. Pure polymers: ● (PSF = a polysulphone [23]; COP = a copolyester [26]; remaining abbreviations as in the text). Miscible polymer blends: ○ PPO-PS [65]; ⊙ PC-COP [26]

in monodisperse PS-CO_2 [28] and (ii) densification (by sub-T_g annealing) [22] or semipermanent dilation (by conditioning in high pressure CO_2) [20] of the polymer in the PC-CO_2 system, can also be successfully interpreted by the above correlation. Assuming the whole of v_E to be included in the microregions acting as specific sorption sites and neglecting any swelling induced by the penetrant, the effective molar volume of CO_2 sorbed at these sites, $V = v_E/s_0$ (where s_0 is in mol cm^{-3}), turns out to be very similar to that in the liquid or liquid solution state or in 4A and 5A zeolite cavities saturated with CO_2 [12,13,20,24]. This is strong evidence in favour of (i) a "hole filling" sorption mechanism analogous to interstitial sorption in zeolites and (ii) the usefulness of s_0 as a measure of sorption capacity for CO_2 (and presumably other condensable gases too).

On the other hand, it is noteworthy that progressively lower s_0 values are found for lighter gases. In fact, it appears that s_0, like K_2, tends to be correlated with the Lennard-Jones parameter of the penetrant gas [39], as illustrated in Fig. 3. Such behaviour is expected for K_2 (cf. the behaviour of K_1 above), but not for s_0, according to the Langmuir localised sorption concept.

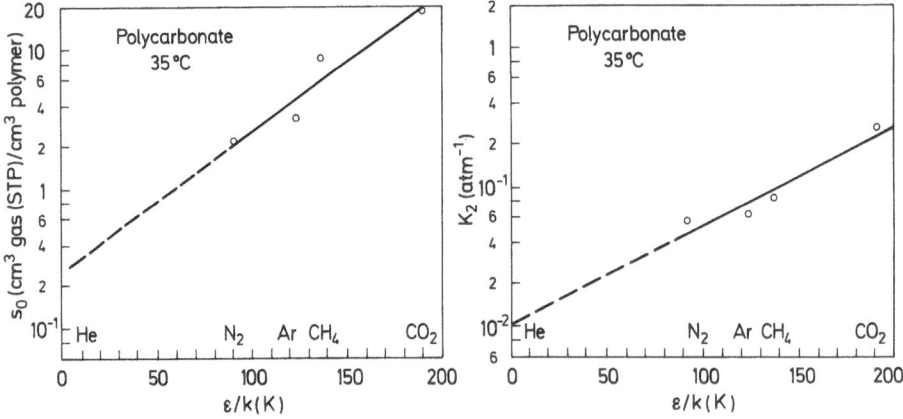

Fig. 3a and b. Correlation of the experimental Langmuir capacity (s_0) and affinity (K_2) parameters in PC at 35 °C with the Lennard-Jones parameter ε/k of the penetrant gas [39]

This concept also requires that there be no polymer — penetrant interaction to such an extent that significant polymer structural changes affecting s_0 and/or K_2 occur. Over the experimental gas pressure range normally used ($p \sim 0-20$ atm), this appears to be so in the case of permanent gases, like N_2; but not in the case of condensable gases, like CO_2. Two effects may be noted here:

First, conditioning of the polymer membrane in high pressure gas causes some permanent dilation, which entails an increase of v_E, and hence of s_0 (both for the conditioning gas and for other gases [19,20]). This effect has already been referred to above in contradistinction to sub-T_g annealing. Upon removal of the conditioning penetrant, the polymer will tend to densify again [40,41]; but, under the conditions of the work described here, the rate of such structural changes is normally sufficiently low to be disregarded. Accordingly, preconditioning the polymer at the highest gas pressure to be employed in a given sorption isotherm determination has now become standard practice, ensuring that the isotherm in question pertains to a relatively stable polymer structure [12,18,20]. It has been further suggested [10] that the preconditioning treatment, in addition to the above general dilation effect, also leads to semipermanent redistribution of the excess free volume among individual microregions of the polymer in a manner favouring the most efficient accomodation of the particular penetrant gas involved in the conditioning process. The idea that (i) during conditioning, the penetrant molecules act effectively as "templates" for the free volume distribution in the polymer and (ii) the resulting free volume distribution remains at least partly frozen-in upon removal of the conditioning penetrant, has been advanced previously in another connection [42] and could go part of the way in explaining the variations of s_0 illustrated in Fig. 3. However, a proper physical theory within the bounds of the Langmuir sorption mechanism leading to a systematic variation of s_0 with the Lennard-Jones parameter of the penetrant gas (analogous to the successful explanation of the temperature dependence of s_0 discussed above) is, at present, lacking.

The second effect to be noted is the (reversible) swelling of the polymer caused by sorption of substantial amounts of penetrant. Whether this also leads to some significant increase of v_E (and hence of s_0) is not clear. On the other hand, it may reasonably be expected, on the basis of Eq. (10), that the concurrent plasticizing action of the penetrant [20,43] on the polymer should cause v_E to diminish. For a clear demonstration of this effect, one must turn to polymer-vapour systems. The sorption isotherms of vinyl chloride in PVC [32,44], illustrated in Fig. 4, provide a good example: the sorption data for T above the T_g of the polymer, as well as those at the higher vapour pressures for T below the T_g of the polymer, are well described [44] by a single curve conforming to an ideal Flory-Huggins isotherm

$$\ln (p_A/p_A^0) = \ln v_A + 1 - v_A + \chi(1 - v_A)^2 \tag{12}$$

where p_A/p_A^0 is the relative vapour pressure of penetrant, v_A is its volume fraction in the membrane and $\chi = 0.98$. Here, the curve given by Eq. (12) should be identified with sorption in the polymer matrix. On this basis, we see that "excess sorption" (and hence the value of s_0) is progressively reduced at the higher penetrant concentrations C_A and finally vanishes at values of C_A (or v_A) at which T_g for the polymer-penetrant mixture attains the experimental temperature [32]. Between the extreme case of vanishing s_0 noted here and the other extreme case of perfectly constant s_0 (assumed in the Langmuir isotherm), intermediate types of behaviour (i.e. a tendency of s_0 to vary to a greater or lesser extent from one end of the isotherm to the other) can obviously be expected. The possibility of significant plasticization effects in the case of CO_2, which is of particular interest here, has been put forward [20,43]; but further work is required to establish at what point in various systems the variability

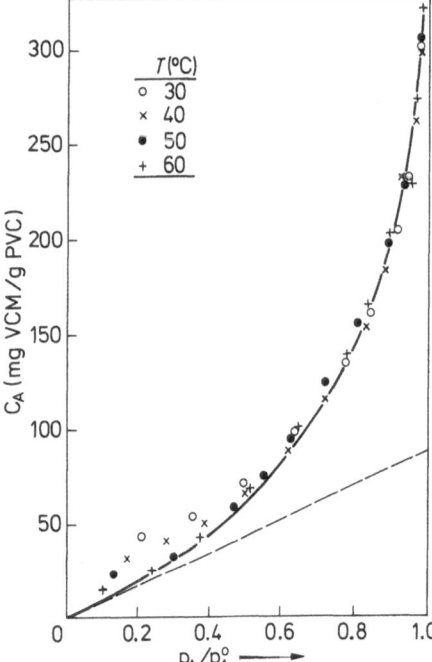

Fig. 4. Experimental sorption isotherms of vinyl chloride (VCM) in PVC in comparison with the calculated curve from Eq. (12) with $\chi = 0.98$ [44]

of s_0 becomes sufficiently pronounced to make the Langmuir-like representation of excess sorption unacceptable. In the case of vapours, on the other hand, there is real difficulty in devising a satisfactory general treatment. Accordingly, some authors have confined themselves to low v_A values [45-47], whilst others have resorted to empirical generalizations of Eq. (8) [31,48,49]. A gas sorption theory based exclusively on plasticization effects has also been proposed [50].

The above review indicates that the model of Eqs. (8) and (9) provides a physically meaningful, though not rigorous, description of gas sorption in glassy polymers. Further clarification of the limitations of the model is desirable. On the other hand, considerable progress has been made in developing this approach as a predictive tool by means of the correlations between model parameters and gas or polymer properties indicated above. Further very recent developments along these lines include correlations between the cohesive energy density of the polymer with K_1, K_2 for CO_2 at 35 °C [51].

3.2 Diffusion of Single Gases

Application of the dual mode concept to gas diffusion in glassy polymers was originally subject to the limitation that $D_{T2} = 0$ in Eq. (6) ("total immobilization model")[6]. Later this simplifying assumption was shown to be unnecessary, provided that suitable methods of data analysis were used [52]. Physically, the assumption $D_{T2} = 0$ is unrealistic, although it is expected that $D_{T2} < D_{T1}$ [52]. Hence, this more general approach is often referred to as the "partial immobilization model".

Treatment of the data is greatly facilitated by considering the P(a) function in preference to the more complicated D(C) [27,52] one. The former function follows simply from Eqs. (6), (8) and (9):

$$P(a) = K_1 D_{T1} + s_0 K_2 D_{T2}/(1 + K_2 a) \tag{13}$$

or, in integral form,

$$\bar{P}(a_0, a_1) = K_1 D_{T1} + [s_0 D_{T2}/(a_0 - a_1)] \ln [(1 + K_2 a_0)/(1 + K_2 a_1)] \tag{14}$$

The integral permeability coefficient $\bar{P}$ may be determined directly from permeation steady-state flux measurements or indirectly from sorption kinetic measurements [27,52]; activity is usually replaced by gas concentration or pressure (unless the gas deviates substantially from ideal behaviour and it is desired to allow for this); and a_0, a_1 (p_0, p_1) are the boundary high and low activities (pressures) respectively in a permeation experiment, or the final (initial) and initial (final) activities (pressures) respectively in an absorption (desorption) experiment.

If the above treatment is formulated in terms of "Fick" diffusion coefficients D_1, D_2 [53] (cf. the introductory section for the difference between a "thermodynamic" and a "Fick" diffusion coefficient), the expressions equivalent to Eqs. (13) and (14) are respectively

$$P(a) = D_1 \frac{dC_1}{da} + D_2 \frac{dC_2}{da} = K_1 D_1 + s_0 K_2 D_2/(1 + K_2 a)^2 \tag{13a}$$

$$\bar{P}(a_0, a_1) = K_1 D_1 + s_0 K_2 D_2/[(1 + K_2 a_0)(1 + K_2 a_1)] \tag{14a}$$

The alternative formulations of the model represented by Eqs. (13), (14), on one hand, and by Eqs. (13a), (14a), on the other hand, do not differ in essence (although the former may be considered preferable from the theoretical point of view). They both predict linear plots of P vs a suitable function of a, if the diffusion coefficients characterizing each sorption mode (D_{T1}, D_{T2} or D_1, D_2) are constant. Eqs. (14) and (14a) were first applied to measurements of the permeability of CO_2 through PC membranes [54]. Plots of $\bar{P}$ (p_0, $p_1 = 0$) vs either $\ln [(1 + K_2 p_0)/p_0]$ or $(1 + K_2 p_0)^{-1}$ gave about equally good linear fits to the data; hence, on practical grounds, there was little to choose among Eqs. (14) and (14a). Values of D_{T1} or D_1 and D_{T2} or D_2 were deduced from the intercept and slope respectively of the appropriate linear plot (knowing K_1, K_2 and s_0 from sorption measurements). D_{T1}, D_1 were found to differ somewhat; but D_{T2}/D_{T1}, D_2/D_1 were practically the same.

Most experimental determinations of D_1, D_2 reported to date [13,15,16,23-26,55,56] have been obtained by the method described above; but sorption kinetics [16,17,27] and permeation time lag measurements [29,57,58] have also been employed for this purpose.

The time lag is given in terms of $P(a)$ by [52]

$$L(a_0, a_1) = l^2 \{\bar{P}(a_0, a_1)(a_0 - a_1)\}^{-3} \times$$
$$\times \int_{a_1}^{a_0} \{a\, S(a) - a_1 S(a_1)\}\, P(a)\, \bar{P}(a, a_1)(a_0 - a)\, da \qquad (15)$$

The expressions resulting by substitution into Eq. (15) from Eqs. (5), (6), (8), (9) and either (13), (14) or (13a), (14a) have been worked out in detail [53], but are too lengthy to quote here. Time lags calculated by means of these expressions, on the basis of the relevant solubility and permeability data, do not always agree with the corresponding experimental values [15,20,39,54,55]. These deviations are attributable to varying degrees of departure from normal "Fickian" diffusion kinetics [1,4,52], depending on the nature of the polymer [15,39,55] and penetrant [39], the temperature [55] and the mode of high gas pressure preconditioning of the membrane [20]. Some evidence of "non-Fickian anomalies" has also been discerned in sorption kinetic measurements [16,17,27]. Consequently, due caution should be exercised when the estimation of D_1, D_2 is based on transient diffusion measurements [52].

So far, it appears that the gas transport properties of glassy polymer membranes, manifested in a decreasing $P(a)$, or increasing $D(C)$, function can be adequately represented by the above dual diffusion model with constant diffusion coefficients D_1, D_2 (or D_{T1}, D_{T2}). We now consider the implications of this model from the physical point of view:

The expectation of $D_2 < D_1$ mentioned earlier arises from the physical picture of low density microregions dispersed in the polymer matrix and largely isolated from one another. (The situation would be quite different if the microregions in question tended to form continuous pathways or channels through the polymer matrix [59]). Thus, all penetrant molecules effectively migrate through the polymer matrix. This process will be characterized by a certain activation energy; but mobilization of penetrant molecules sorbed at the specific sites will require additional activation energy for prior desorption into the polymer matrix. In terms of the Eyring theory [60], we have

$$\frac{D_2}{D_1} = \frac{\lambda_2}{\lambda_1} \exp\left(\frac{\overline{\Delta S_2^{\ddagger}} - \overline{\Delta S_1^{\ddagger}}}{R}\right) \exp\left(-\frac{\overline{\Delta H_2^{\ddagger}} - \overline{\Delta H_1^{\ddagger}}}{RT}\right) \qquad (16)$$

where λ_1, λ_2 represent mean jump length, $\overline{\Delta S_1^{\ddagger}}$, $\overline{\Delta S_2^{\ddagger}}$ activation entropy, and $\overline{\Delta H_1^{\ddagger}}$, $\overline{\Delta H_2^{\ddagger}}$ activation enthalpy, of penetrant molecules sorbed in the matrix and at specific sites respectively. On the basis of the above mechanism of diffusion, it may be expected [61] that $\lambda_1 \simeq \lambda_2$ and that the magnitude of $\overline{\Delta S_2^{\ddagger}} - \overline{\Delta S_1^{\ddagger}}$, $\overline{\Delta H_2^{\ddagger}} - \overline{\Delta H_1^{\ddagger}}$ will parallel that of the corresponding differences in the entropies and enthalpies of sorption. Since [60] the trend in D is governed primarily by that in $\overline{\Delta H^{\ddagger}}$, it follows that

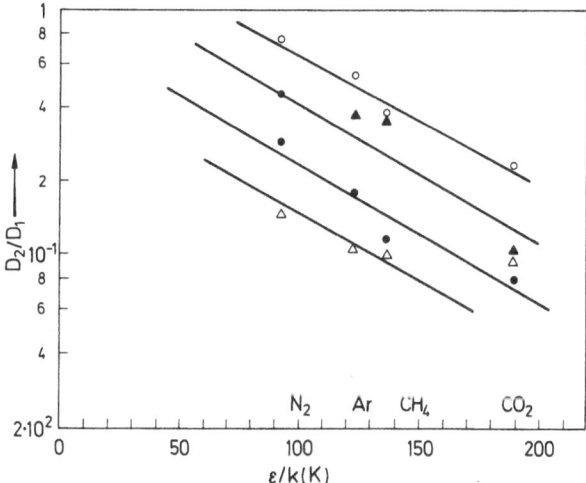

Fig. 5. Correlation between experimental D_2/D_1 values with the Lennard-Jones parameter of the penetrant gas for four glassy polymers [25]: ○ a copolyester [26]; ▲ a polysulphone [23]; ● PC; △ PPO

D_2/D_1 would be expected not only to be <1, but also to decrease as the penetrant becomes more strongly sorbed [61]. On the other hand, the absolute values of D_1, D_2 should obviously be affected in a similar manner by factors such as the size or shape of the penetrant molecule and the cohesive energy density or free volume of the polymer matrix.

The data obtained so far refer to a variety of polymers (including PET [13,55,57], PAN [15], PC [20,27,39,54,56], PS [27,57], PVAc [29], EC [24], PPO [25], a polyimide [16,17] a copolyester [26] and a polysulphone [23]) and gaseous penetrants (the most common ones being CO_2, N_2, Ar, CH_4 and C_3H_8). In accordance with the above theoretical expectations, D_2/D_1 is <1 and varies inversely with the Lennard-Jones parameter ε/k characteristic of the gas (cf. Fig. 5) [25]. Slight increases or decreases of D_2/D_1 for CO_2 in PC have been reported upon sub-T_g annealing [56] or high pressure CO_2 conditioning [20] respectively. (These results would, incidentally, be understandable if the sites created by conditioning or destroyed by annealing tend to be more "energetic" than the others). Reported activation energy values [15,16,27,55] do not always [27] conform to the requirement $\overline{\Delta H}_1^{\ddagger} < \overline{\Delta H}_2^{\ddagger}$, although the situation is more satisfactory than in the case of the corresponding enthalpies of sorption discussed above (despite the fact that the determination of D_2 is also subject to considerable experimental error). Much attention has been given to the observation that $\overline{\Delta H}_1^{\ddagger}$ usually turns out to be less than the value above T_g [8,55] and an interpretation in terms of the excess free volume of the glassy polymer has been given [62]. This interpretation can be reconciled with the present model by assuming that the polymer matrix of the glassy polymer is not "fully densified" (i.e. not all of the excess free volume is in the specific sorption site regions).

The dependence of D_1 on gas molecular size has been found to be He $> CO_2$ $>$ Ar $> N_2 > CH_4$ in four different polymers [23,25,26,39]. This trend correlates smoothly with the "minimum effective gas molecular diameter" deduced from molecular sieving effects in zeolites [39]. The corresponding trend of D_2 is not so clearcut, but the uncertainty of D_2 values should be borne in mind. The fact that there is a general positive correlation between D_1 and D_2 in the above systems, when either the gas or the polymer is varied, is illustrated by Fig. 6 [25].

The above results indicate that the general characteristics of gaseous diffusion in glassy polymer membranes can be represented reasonably well in terms of the dual-mode concept. The basic reason for the observed increasing D(C) function is seen to be the concurrent increasing proportion of less strongly sorbed (and hence more easily activated) penetrant molecules. The model is, no doubt, highly idealized, but is nevertheless shown to be physically reasonable and consistent with the correspond-

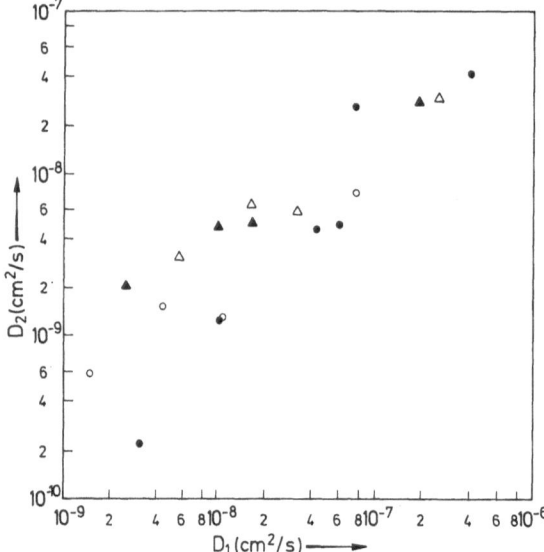

Fig. 6. Correlation between experimental D_1 and D_2 values for the same polymer-gas systems as in Fig. 5: ● CO_2; △ Ar; ▲ N_2; ○ CH_4 [25]

ing sorption treatment. At the present time, it does not appear that an alternative treatment, which seeks to attribute the variation of D(C) wholly to plasticization effects [34,63] can lay claim to comparable status. Although plasticization effects are not unexpected in the case of the more strongly sorbed gases (cf. the preceding subsection), no significant deviations requiring modification of the assumption of constant D_1 and D_2 have been discerned in the data so far accumulated (although some preliminary data for CA-CO_2, which do not appear to fit into this pattern have recently been reported [63]). On the other hand, marked plasticization effects can be expected and have indeed been observed [47-49] in the case of vapours. These have been dealt with by an empirical extension of the dual mode diffusion model (involving a D_1 exponentially dependent on the concentration of mobile penetrant with D_1/D_2 = const.) [64].

Application of the dual mode sorption and diffusion models to homogeneous polymer blend-gas systems [26,65] and filled polymers [66] has also been reported.

3.3 Gas Mixtures

The sorption and diffusion behaviour of gas mixtures is of particular interest from the point of view of membrane gas separation, which is steadily gaining in importance by virtue of its low energy requirements. On the basis of the dual mode sorption model, one may reasonably expect that sorption of a binary gas mixture A, B in the polymer matrix will exhibit little gas-gas interaction and hence will tend to occur essentially additively. In the Langmuir-like mode of sorption, on the other hand, there will be competition between A and B for the limited number of available sites. These considerations led [67] to the following reformulation of Eqs. (8) and (9)

$$S_{A1} = K_{A1} \tag{17}$$

$$S_{A2} = S_{A0}K_{A2}/(1 + K_{A2}a_A + K_{B2}a_B) \tag{18}$$

for component A and analogous expressions for B; where the constants $K_{A1}(K_{B1})$, $K_{A2}(K_{B2})$ and $S_{A0}(S_{B0})$ are determined from the sorption isotherms of the respective single gases.

The permeability functions $P_A(a_A)$, $P_B(a_B)$ may be derived from either Eq. (13) or Eq. (13a) [68]:

(a) Eq. (13) is evidently transformed to

$$P_A(a_A) = K_{A1}D_{TA1} + S_{A0}K_{A2}D_{TA2}/(1 + K_{A2}a_A + K_{B2}a_B) \tag{19}$$

To obtain $\bar{P}(a_{A0}, a_{A1})$ for a steady state permeation experiment, we must have a relation between $a_A(X)$ and $a_B(X)$. For gases of similar sorption and diffusion properties and $a_{A0} \sim a_{B0}$, $a_{A1} \sim a_{B1}$

$$(a_A - a_{A1})/(a_{A0} - a_{A1}) = (a_B \quad a_{B1})/(a_{B0} - a_{B1})$$

Hence, substituting in Eq. (19) and integrating, we find

$$\bar{P}_A(a_{A0}, a_{A1}) = K_{A1}D_{TA1} + \frac{S_{A0}D_{TA2}}{(a_{A0} - a_{A1})} \ln \frac{1 + K_{A3} + K_{A4}a_{A0}}{1 + K_{A3} + K_{A4}a_{A1}} \tag{20}$$

where

$$K_{A3} = K_{B2}\{a_{B1} - a_{A1}(a_{B0} - a_{B1})/(a_{A0} - a_{A1})\}$$

$$K_{A4} = K_{A2} + K_{B2}(a_{B0} - a_{B1})/(a_{AB0} - a_{A1})$$

(b) Eq. (13a) becomes

$$P_A(a_A) = K_{A1}D_{A1} + S_{A0}K_{A2}(1 + K_{A3})D_{A2}/(1 + K_{A3} + K_{A4}a_A)^2 \tag{19a}$$

and the integral steady-state permeability under the same conditions as those envisaged for Eq. (20) is

$$\bar{P}_A(a_{A0}, a_{AI}) = K_{AI}D_{AI} + \frac{S_{A0}K_{A2}(1 + K_{A3})D_{A2}}{(1 + K_{A3} + K_{A4}a_{A0})(1 + K_{A3} + K_{A4}a_{AI})}$$

(20a)

Because of non-adherence of the site sorption mode to a strict Langmuir mechanism, as noted previously, Eq. (18), as well as Eqs. (20) or (20a), must, at the quantitative level, be validated experimentally. This can be done most conveniently by varying the partial pressure of one component at various constant partial pressures of the other. Sorption data of this type have recently been reported for PMMA-CO$_2$, C$_2$H$_4$ at 35 °C [69,70]. As shown in Fig. 7, the agreement between experiment and calculation from the pure component isotherms, though not perfect, is nevertheless quite impressive.

Analogous data for permeation are also of great interest. The simple treatment discussed above predicts that the polymer matrix permeability component $P_{AI} = S_{AI}D_{TAI}$ will be essentially unaffected by the presence of B. This should be true of D_{TA2} too, since this parameter also describes diffusion through the polymer matrix (cf. preceding subsection). However, $P_{A2} = S_{A2}D_{TA2}$ should tend to diminish by virtue of the effect of B on S_{A2} in accordance with Eq. (18) [68]. In practice, it is usual to have $a_{AI} = a_{BI} = 0$. Then, $K_{A3} = 0$ and $K_{A4} = K_{A2} + K_{B2}a_{B0}/a_{A0}$. Thus, if a_{A0} is varied with either $a_{B0} = $ const. or $a_{B0}/a_{A0} = $ const., Eqs. (20) or Eq. (20a) should yield straight lines when plotted in the same manner as Eqs. (14) or (14a)

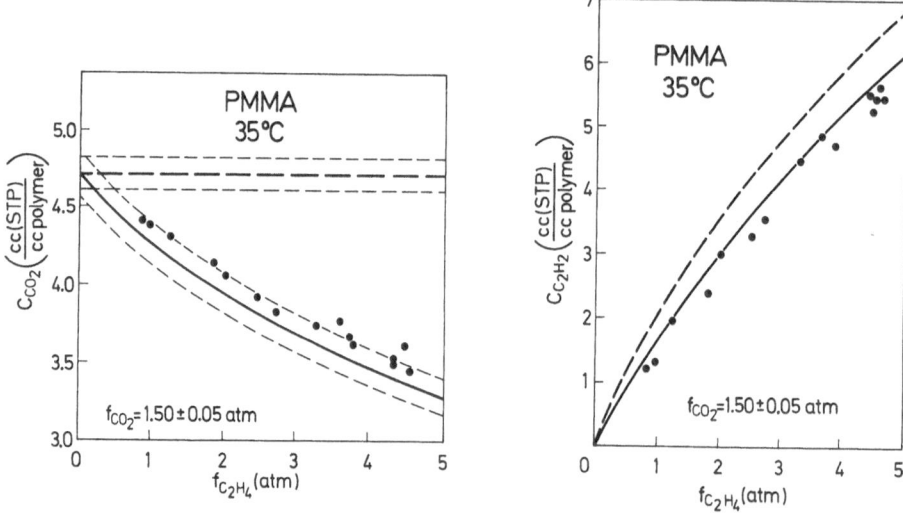

Fig. 7. Mixed gas sorption isotherms for PMMA-CO$_2$, C$_2$H$_4$ in terms of fugacities f, at f$_{CO_2}$ = 1.50 ± 0.05 atm, T = 35 °C. Experimental data (●) in comparison with lines calculated from pure component isotherms on the basis of (i) additive sorption (– – –), or (ii) the treatment of Eqs. (17), (18) (———); - - - - - - - - maximum range of model predictions due to ±0.05 variation in f$_{CO_2}$ [69]

(with $a_1 = 0$) respectively. Appropriate $\bar{P}_A(a_{A0})$ data have recently become available for PC-CO_2 at a fixed upstream pressure of isopentane (cf. Fig. 8) [71], for "Kapton" polyimide-CO_2 at two different upstream water vapour pressures [72] and for "Kapton"-CO_2, CH_4 at fixed upstream gas mixture composition [73]. In all cases the predicted depression of the permeability of one penetrant by the other is evident. Furthermore, the quantitative predictions of Eq. (20a) from pure component behaviour appear to be reasonably successful in the case of the gas mixtures [71,73], in spite of the simplifying assumption inherent in this equation (v.s.). Reliable prediction of the effect of water vapour on $\bar{P}_{CO_2}$ in "Kapton" was not possible, because of lack of information on water sorption. (The effect was found to be considerably larger than expected if K_{B2} were equated to K_2 for PAN-H_2O and it is suggested [72] that the difference could be due to antiplasticization of "Kapton" by water).

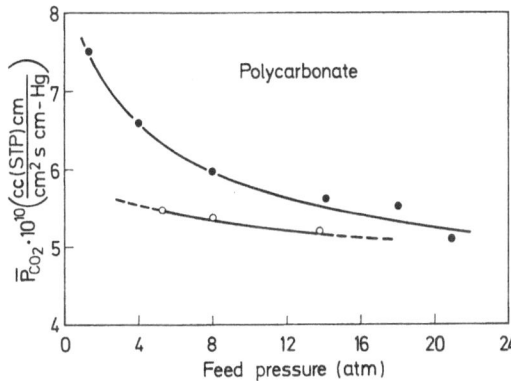

Fig. 8. Experimental integral permeability data for PC-CO_2: ● pure CO_2; ○ in the presence of 117.8 torr of isopentane on the upstream side ($p_1 = 0$) [71]

The general conclusion to be drawn from the above discussion is that the dual gas sorption and mobility model has very useful predictive potential in the field of membrane gas separation. At the present stage, no comparable predictive capability is discernible in the alternative treatments so far published [33,34,63].

4 Dual-Mode Ionic Sorption and Diffusion in Charged Polymer Membranes

4.1 High Affinity Ionic Species

It is particularly interesting and instructive to note that application of Henry + Langmuir dual-mode sorption and diffusion models is not confined to glassy polymer-gas systems. Sorption and transport of high affinity ionic species, exemplified by anionic dyes, in charged polymers, exemplified by polyamides at low pH, has been treated in the same way. These systems are of considerable importance both from the biomimetic and from the textile processing point of view, but have received limited attention in previous reviews of dual mode sorption. Accordingly, a more thorough coverage is attempted here (cf. also a review in Japanese [74]).

The uptake of anionic dyes from acidified solutions by polyamides has traditionally been treated as Langmuir sorption on sites usually identified with the terminal amino-groups of the polymer [75,76]. It was also early realized that additional sorption can occur at sufficiently low pH, known as "overdyeing" [75,76].

The simultaneous operation of these sorption modes was first shown clearly in the work of Atherton et al. [77], who measured sorption isotherms of various anionic dyes in Nylon at a pH of 3.2. At this pH, the polymer is essentially in the form $(NH_3^+ - Ny - CO_2H)A^-$, where "Ny" is the Nylon chain apart from the end groups and HA is the simple acid used for acidification of the solution. The sorption of dye then follows an initially convex-upward isotherm, which tends to become linear at higher concentrations of dye in solution C_{DS}. Extrapolation of the linear part of the isotherm to $C_{DS} = 0$ generally gave sorbed dye concentrations C_D approximately equal to s_B/z, where s_B is the concentration of amine end groups in the polymer and z is the valence of the dye anion (D^{z-}). These findings are consistent with an $-NH_3^+$ site sorption mode with each dye ion occupying z sites, coupled with a Henry law, or (more properly in this field) Nernst, mode of non-specific sorption (or, equivalently, of sorption on amide groups of which there is a large number) probably involving the undissociated dye acid (H_zD) [77]. Further evidence for the existence of the latter "overdyeing" mode was obtained by varying s_B [78]: the sorbed dye concentration at pH = 3 and constant C_{DS} was found to be an essentially linear function of s_B extrapolating to $C_D > 0$ at $s_B = 0$. The "overdyeing" mode becomes relatively more important as (i) the dye-bath pH decreases, (ii) s_B is reduced, (iii) the valence of the dye anion becomes lower or its affinity for the polymer higher; and it may be represented by Eq. (8), or, setting a $\simeq C_{DS}$, by

$$C_{D1}/C_{DS} \simeq S_1 = K_1 = \text{const.} \qquad (21)$$

in accordance with what has been said above; although detailed tests of Eq. (21) have not been published.

Similarly, the site sorption mode under the conditions described above is represented by Eq. (9), or setting a $\simeq C_{DS}$, by

$$\frac{C_{D2}}{C_{DS}} \simeq S_{D2} \simeq \frac{s_0 K_2}{z(1 + K_2 C_{DS})} \quad \text{or} \quad C_{DS} \simeq \frac{\vartheta_D}{K_2(1 - \vartheta_D)} \qquad (22)$$

where $\vartheta_D = zC_{D2}/s_0$ denotes the fraction of sites occupied by sorbed dye and $s_0 \simeq s_B$. However, a true Langmuir sorption mechanism is difficult to justify physically. The nearly stoichiometric relation between s_0 and s_B, in particular, is difficult to reconcile with the physical occupation of sites in a fixed lattice when z = 2 or 3. Hence the use of Eq. (22) for z > 1 was justified on essentially empirical grounds [77].

A physically more realistic approach has been developed [79,80] by considering the ion exchange equilibrium

$$z(-NH_3^+ A^-) + D^{z-} \rightleftharpoons (-NH_3^+)_z D^{z-} + zA^-$$

in a dyebath containing a mixture of acids H_zD and HA at constant total acid concentration C_{HS}^0. The equilibrium relation

$$K \simeq C_{D2}C_{AS}^z/C_{DS}C_A^z$$

in conjunction with the electroneutrality conditions

$$C_{AS} + zC_{DS} = C_{HS}^0 ; \qquad C_A + zC_{D2} = s_B$$

yields [80]

$$K \simeq \frac{C_{D2}(C_{HS}^0 - zC_{DS})^z}{C_{DS}(s_B - zC_{D2})^z} \simeq \frac{(C_{HS}^0 - zC_{DS})^z \, \vartheta_D}{C_{DS}zs_B^{z-1}(1 - \vartheta_D)^z} \tag{23}$$

For $z = 1$, Eq. (23) reduces to Eq. (22) with

$$K_2 = (K - 1)/C_{HS}^0 ; \qquad s_0 = Ks_B/(K - 1) \tag{24}$$

showing that, according to this treatment, (i) K_2 is an apparent Langmuir affinity constant dependent on C_{HS}^0, and (ii) s_0 is an apparent Langmuir capacity constant exceeding s_B, unless $K \gg 1$. For higher values of z, solutions of Eq. (23) become increasingly complicated. However, for $K \gg 1$, $zC_{DS} \ll C_{HS}^0$ over most of the range in C_{DS} covered by the isotherm and Eq. (23) simplifies to [80]

$$C_{DS} \simeq \vartheta_D/K_2(1 - \vartheta_D)^z ; \qquad K_2 = zKs_B^{z-1}(C_{HS}^0)^{-z} \tag{23a}$$

indicating a definite departure from the Langmuir form when z > 1.

Application of the dual sorption models embodied in Eq. (5) in conjunction with Eqs. (21) and either (22) or (23) has, so far been limited by the shortage of appropriate data. The most satisfactory data in this respect relate to the sorption of a monobasic dye (Orange II or C.I. Acid Orange 7, referred to hereafter as Dye I) by undrawn [81] or biaxially drawn [82] Nylon 6 film. The resulting parameter values are summarized in Table 2. Data on a divalent and a trivalent dye (C.I. Food Yellow 3 and sulphanilic acid-R acid, referred to hereafter as Dyes II and III respectively) are less useful, because they cover only the upper C_D range [82]. Clearly, more extensive sorption data are highly desirable. For the moment, it is worth noting in Table 2 that $s_0 \simeq s_B$ for $K \gg 1$, whereas $s_0 > s_B$ for $K \sim 1$, in accordance with the prediction of Eq. (24). It has also been noted that fitting the sorption data of Dye II by Eq. (23a) leads to an s_0 value in better accord with s_B than was obtained with the aid of Eq. (22) [83].

As already pointed out in the preceding section, diffusion may be treated in terms of either of the functions $P(a)$ or $D(C)$. The former function is much simpler and has been used in the case of glassy polymer-gas systems for this reason, as we have seen. Its application to the present systems is more recent [83,84]. Eqs. (13) and (14) {or Eqs. (13a) and (14a)} are directly applicable to monovalent dye anions. The same equations may be applied to multivalent dye ions following the (empirical) treatment of Eq. (22), provided that s_0 in Eqs. (13) or (14) is replaced by s_0/z. Fig. 9 shows that

Table 2. Sorption of the monovalent anionic dye Orange II (Dye I) by different Nylon 6 membranes at constant dyebath pH

T °C	pH	K_1	K_2	s_0 mol/Kg		s_B	K	A^-	Ref.
		dm³/kg	dm³/mol	(a)	(b)	mol/kg			
90	3.6	1.5	0.12×10^5	0.054	0.060	0.045	4.0	CH_3COO^-	[81]
50	2.2	160	6.4×10^5	0.037	0.040	0.040	4×10^3	Cl^-	[82]

Notes: (a) s_0 values obtained by fitting Eq. (22) to the experimental isotherm;
(b) s_0 values predicted by Eq. (24)

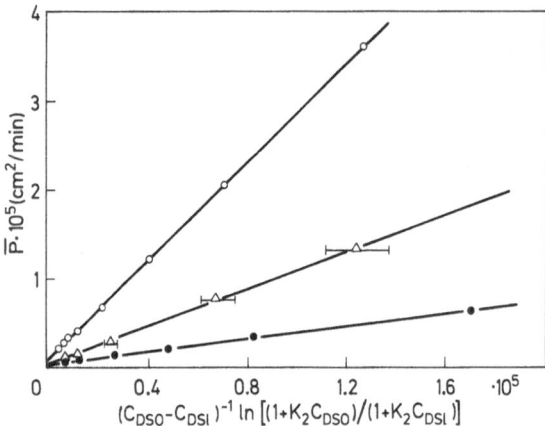

Fig. 9. Experimental integral permeability plots for Dyes I–III through Nylon 6 membranes according to Eq. (14) [84] (with C_{DSI} varying between zero and a small finite value to the extent indicated by the horizontal bars): ○ Dye I; △ Dye II; □ Dye III

reasonably good linear plots have indeed been obtained for $\bar{P}(C_{DSO}, C_{DSI})$ vs the proper function of the boundary concentrations C_{DSO}, C_{DSI}, in the case of all three Dyes I–III [84]. According to the treatment of Eqs. (23) or (23a), on the other hand, the expression for S_{D2} which must be inserted in Eq. (6) becomes increasingly complex and unwieldy for $z > 1$, as already noted. Nevertheless, this treatment has now been applied to $\bar{P}(C_{DSO}, C_{DSI})$ for Dye II with results at least as good as those shown in Fig. 9 [83].

In the systems examined here, the site sorption mode is the major component in both sorption and permeation. As a result we often have $C_D \simeq C_{D2}$ over most of the isotherm. This greatly simplifies the D(C) function; which, in practice, is often derived directly from measurements of diffusion profiles [81,82]. Under these conditions, analytical expressions for D(C) can be derived in terms of $\vartheta_D \simeq zC_D/s_0$ even from Eq. (23a) [80]. Thus, from Eqs. (4), (5), (6), (21) and either (22) or (23a), we obtain respectively

$$D = \{D_{T1} + \beta D_{T2}(1 - \vartheta_D)\}/\{1 + \beta(1 - \vartheta_D)^2\} \qquad (25)$$

$$D = \{D_{T1} + \beta D_{T2}(1 - \vartheta_D)^z\}/[1 + \beta(1 - \vartheta_D)^{z+1}/\{1 + (z - 1)\vartheta_D\}] \,(26)$$

where

$$\beta = s_0 K_2 / z K_1$$

Eqs. (25) and (26) are, of course, identical when $z = 1$ and have been applied successfully to Dye I in two cases [81,82] (cf. Fig. 10). Eq. (26) has been applied with reasonable success to two divalent dyes [80]. Finally, Eq. (25) was tested against data for Dyes II and III. Some discrepancies between the experimental and calculated $D(C)$ curves have been noted at the high concentration end [82]. No comparable anomalies are noticeable in the corresponding steady-state permeation data plotted in Fig. 9 (but it should be borne in mind that the high C region tends to be deemphasized in the latter type of plot). Further data are desirable.

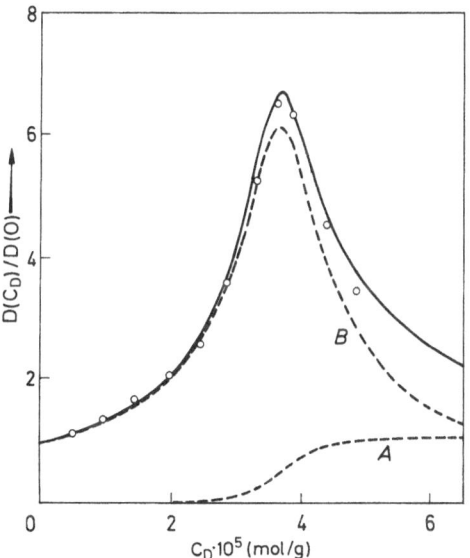

Fig. 10. Concentration dependence of the diffusion coefficient of Dye I in Nylon 6 determined from diffusion profiles (film-roll method), in comparison with theoretical calculations from Eq. (25) with $D_{T2} = 1.5 \times 10^{-10}$ cm^2/s and $D_{T2}/D_{T1} = 0.7$ (full line) [82,84]. The individual contributions of the first and second terms of Eq. (25) are shown by the broken lines A and B respectively

The values of D_{T2} deduced from the diffusion profile and permeation data are compared in Table 3. Quantitative agreement is not very close and there is need to assess the reliability of these methods more fully. The general pattern of the results, however, is the same: D_{T2} is inversely correlated with K_2, i.e. with the apparent affinity of the dye for the polymer. Dyes I–III differ only in the number of $-SO_3^-$ groups; it will be noted that the respective variations in K_2 and D_{T2} are rather minor. A much larger variation, still in keeping with the general pattern, is found when the Nylon substrate is changed. The relative importance of the "overdyeing" mode is in keeping with the behaviour indicated earlier in this subsection. Values of D_{T1} could not be determined in the above systems with any precision (cf. the large values of β in Table 3); but all available evidence [82,84] (cf. also Fig. 10) points to $D_{T1} \sim D_{T2}$, in quite marked contrast to what is found in the glassy polymer-gas systems discussed in the preceding section. According to the discussion given in that section, this

Table 3. Sorption and diffusion parameters for Dyes I–III in Nylon 6 following the treatment of Eq. (22)

Dye	z	T °C	pH	K_1	$K_2 \times 10^{-5}$	β	$D_{T2} \times 10^{10}$ cm²/s	
				dm³/kg	dm³/mol		(a)	(b)
I [81]	1	90	3.6	1.5	0.12	430	67	—
I [82, 84]	1	50	2.2	160	6.4	150	1.5_5	1.0
II [82,84]	2	50	2.2	24	11	830	1.4_7	0.78
III [82,84]	3	50	2.2	13	20	1900	0.63	0.40

Notes: (a) D_{T2} determined from transient diffusion profiles by Eq. (25);
(b) D_{T2} determined from sorption and steady-state permeation data by Eq. (14) with s_0 replaced by s_0/z

finding implies that the energy of sorption is not very different for the two species of sorbed dye. Some preliminary data on the effect of NaCl on the sorption and permeation of Dye III have been reported [74].

In summary, the results are so far encouraging, but further data are needed for applying and testing Eq. (23).

4.2 Low Affinity Ionic Species

Application of dual mode sorption models to low affinity ionic species has been reported. Pertinent examples are mentioned here to illustrate the variety of models used:

The sorption isotherm of NaCl in an amphoteric ion exchange resin containing $-NR_4^+$ and $-CO_2^-$ groups, measured after prior washing of the resin with water [85], was well represented by a Langmuir + Nernst dual mode sorption model at salt concentrations not exceeding 0.2 mol dm⁻³. A detailed physical interpretation of the relevant parameters was not given, however, neither was the dual mode concept utilized in a corresponding diffusion study [86].

The sorption of a weak electrolyte by a charged polymer membrane is another case where Nernst + Langmuir-like dual mode sorption, involving the undissociated and dissociated species respectively, may be expected. The concentration of each species in solution follows, of course, from the dissociation constant of the electrolyte. The sorption isotherms of acetic acid and its fluoroderivatives have been analysed in this manner, and the concentration dependence of the diffusion coefficient of acetic acid interpreted resonably successfully, using Nylon 6 as the polymer substrate [87]. In this case the major contribution to the overall diffusion coefficient is that of the Nernst species; consequently D_{T2} could not be determined with any precision. By contrast, in the case of HCl, which was also investigated [87], no Nernst sorption or diffusion component could be discerned down to pH = 2; and the overall diffusion coefficient obeyed the relation $D = D_{T2}/(1 - \vartheta_D)$, which is the limiting form of Eq. (25) when β → ∞.

5 Macroscopically Homogeneous Binary Composite Membranes

Binary composite membranes constitute the chief example of membranes classified under (b) in the introductory section. They include binary polymer blends or block or graft copolymers exhibiting a distinct domain structure, filled or semicrystalline polymers and the like.

If the component phases denoted by A and B are of sufficient, but still microscopic, size (cf. introductory section) and do not interact appreciably, their individual sorption and diffusion properties may be deduced from measurements on the pure bulk phases. Then, the overall solubility coefficient is given by an additive relation analogous to Eq. (5), except that the volume fractions v_A, $v_B(= 1 - v_A)$ of the respective components in the membranes must be taken into account:

$$S = v_A S_A + v_B S_B \tag{27}$$

The corresponding evaluation of P is a much more difficult matter, however, even in the case of constant P_A and P_B which will concern us here exclusively. The additive relation analogous to Eq. (6), namely

$$P = v_A P_A + v_B P_B \tag{28}$$

is valid when the component phases are in the form of laminae parallel to the direction of flow. In this "parallel configuration" both phases can be described as "fully continuous" in the sense that all flow lines are straight and lie wholly within either A or B. The resulting value of P turns out to be the maximum possible for any given values of P_A, P_B and v_A. The corresponding minimum value of P results from a "serial configuration" of A and B, where both phases are "fully discontinuous". Here, the aforesaid laminae are oriented normal to the direction of flow and must all be crossed by all flow lines. P is given by the well known formula

$$P^{-1} = v_A P_A^{-1} + v_B P_B^{-1} \tag{29}$$

Between the upper and lower bounds set by Eqs. (28) and (29) for given v_A, P_A and P_B, the value of P will depend on the precise geometrical arrangement of the component phases. The problem is not analytically tractable, however, except in simple or idealised cases. Consequently, a large number of formulae of varying degrees of approximation and different physical connotations have been developed in various fields. The relations best known in the diffusion field appear in reviews by Barrer [88] and Crank [89]; but appreciation of their relative merits and physical significance is as yet very limited. Ideally, one would like to know which formula is appropriate for what type of composite membrane structure or, inversely, to deduce structural information about the membrane from measurements of P as a function of v_A.

Some progress along these lines has recently been made [90] in a comparative study of some of the principal formulae quoted by Barrer [88]. Most of these refer to binary composite materials consisting of phase A dispersed in microparticulate form in a continuous matrix of B. In the study in question [90], attention was first drawn to the upper and lower bounds of Eqs. (28) and (29) already mentioned, in place of the more restrictive ones of de Vries [91], also quoted by Barrer [88]. Another important point [90] relates to the Maxwell equation which is perhaps the simplest formula for a dispersion of A in a continuum of B:

$$\frac{P}{P_B} = 1 + \frac{3v_A}{(P_A/P_B + 2)/(P_A/P_B - 1) - v_A} \tag{30}$$

Eq. (30) is commonly considered to apply to a dilute dispersion of spheres [88,91]. It is now shown to be significant over the whole composition range [90]. Of particular interest is the fact that good agreement was found (within reasonable limits), in the region of $v_A \rightarrow 1$, with the appropriate form of the Barrer-Petropoulos model of cubes [92]. It was concluded that Eq. (30) can, in all likelihood, provide a useful structural reference point over the full v_A range corresponding to a dispersion of isometric (i.e. non-elongated) particles packed so as to be as far from touching as possible [90].

Deviation from this reference configuration, e.g. by introducing some randomness of packing, results, at given v_A, P_A and P_B, in a higher (lower) P if $P_A/P_B > 1$ (<1). Elongated particle shapes also cause P to increase or decrease, according as the long axes of the particles tend to be oriented predominantly along or across the direction of flow. Viewed in this light, the Bruggeman relation [88]

$$(P_A - P)(P_B/P)^{1/3} = (1 - v_A)(P_A - P_B) \tag{31}$$

which yields higher (lower) P values than Eq. (30) when $P_A/P_B > 1$ (<1), may be interpreted as representing a moderate degree of random packing. In the same way, the Böttcher relation [88]

$$(P - P_B)(P_A + 2P)/3P = v_A(P_A - P_B) \tag{32}$$

may be considered to describe random mixing. Here, components A and B are equivalent and either can become a continuous phase, if present in sufficient excess ($v_A > 0.67$ or $v_A < 0.33$ respectively). In the intermediate composition range, each component is partly continuous and phase inversion develops gradually as v_A is varied between 0.33 and 0.67.

Composite polymer-gaseous penetrant systems are the obvious choice for practical application of the above ideas; but suitable data properly treated are rather scarce, as may be ascertained from the relevant reviews [88,93]. Among more recent experimental work of this type, one may note measurements of the permeability of a series of gases in styrene (A) — butadiene (B) — styrene (A) block copolymers, consisting either of rodlike domains A dispersed in a matrix of B (with $v_A = 0.26$) or of alternating lamellae of A and B of varying orientation (with $v_A = 0.37$) [94]. Gas permeabilities of 2-vinyl pyridine (A) — isoprene (B) copolymers exhibiting the latter type of structure have also been measured [95]. In all cases, the experimental overall permeability values P lay within the bounds set by Eqs. (28) and (29). Measurements of S for the styrene-butadiene copolymers over a range of compositions (including blends of the copolymers with pure polystyrene) deviated appreciably from Eq. (27). This points to appreciable interaction between the component phases, which was indicated by other evidence also [96–98] and was attributed to "mixing" of the components at the domain boundaries.

The permeability of a number of gases through acrylonitrile-vinyl acetate (PAN-PVAc) and acrylonitrile-styrene (PAN-PS) copolymers has been studied over a wide range of composition [99]. Calculations on the basis of Eq. (30), or of the (empirical) Higuchi equation [100]

$$\frac{P}{P_B} = 1 + 3v_A \left[\frac{P_A/P_B + 2}{P_A/P_B - 1} - v_A - \frac{0.78(1 - v_A)(P_A/P_B - 1)}{P_A/P_B + 2} \right]^{-1} \tag{33}$$

showed that the observed behaviour can be accounted for semiquantitatively, if PAN (rather than PVAc or PS) is considered to constitute the continuous phase. The lack of quantitative agreement was attributed to interaction between the component phases, varying degree of crystallinity, etc. [99].

A system with clearly defined disperse (A) and continuous (B) component phases is afforded by copolymers of styrene (A) grafted onto a polydimethyl siloxane matrix (B) [101]. Lack of appreciable interaction between the components was indicated by gas solubility and T_g measurements. The permeability coefficient of propane and other paraffins over a composition range $v_A = 0 - 0.55$ followed the trend described by Eqs. (30)–(33) (with $P_A = 0$, in view of the fact that the polystyrene phase is practically impermeable). Of particular relevance to the present discussion is the close agreement with the Bruggeman, and definite deviation from the Böttcher, equations at higher v_A (cf. Fig. 11). Corresponding block copolymer membranes with $v_A = 0.34$ also fitted into this pattern, except in one case where the structure was found to be lamellar and P was considerably lower.

It is particularly interesting to compare the above behaviour with that of the permeability of propane through membranes consisting of polydimethyl siloxane

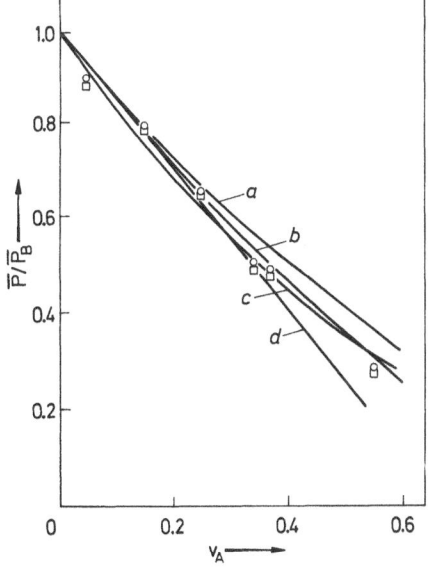

Fig. 11. Relative integral permeability $\bar{P}/\bar{P}_B$ of membranes composed of copolymers of styrene (A) grafted onto a polydimethyl siloxane matrix (B) as a function of composition (volume fraction of A): ○ C_3H_8; □ iso-C_4H_{10}; (a) Eq. (30); (b) Eq. (31); (c) Eq. (33); (d) Eq. (32) [101]

(PDMS)-poly(ethylene-co-propylene) rubber (EPR) blends covering the full composition range [102]. A mechanical blending process can reasonably be expected to correspond to what was termed above "random mixing". Hence, in accordance with the ideas of Ref [90] (cf. the relevant discussion above), the $\bar{P}$ — v_A relation should here tend to follow the Böttcher equation rather than any of the others. Fig. 12 shows that this is indeed so; Eq. (32) is the only one among those considered which can describe the data well over the full composition range. Non-interaction among the component phases was indicated by conformity to Eq. (27), whereas electron microscopy indicated continuous and disperse phase structures at either end, and phase inversion in the middle, of the composition range. In addition to this, earlier data on the transport of N_2 in blends of natural rubber and a butadiene-acrylonitrile copolymer (containing 26% by weight of acrylonitrile) [103] have, also been shown to conform reasonably well to Eq. (32) [102].

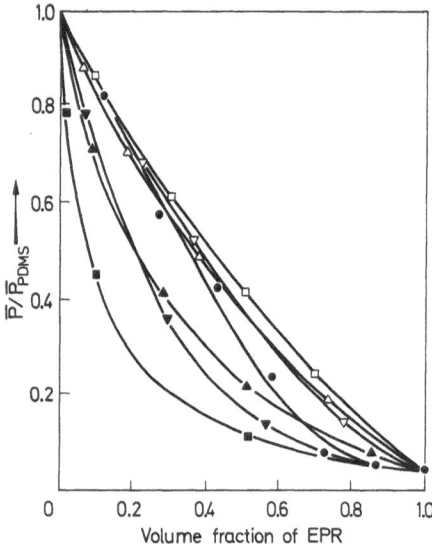

Fig. 12. Relative integral permeability $\bar{P}/\bar{P}_{PDMS}$ of propane through polydimethyl siloxane (PDMS) — poly(ethylene-copropylene) rubber (EPR) blend membranes as a function of blend composition: ● experimental; —— Eq. (32); □, ■ Eq. (30); △, ▲ Eq. (33); ▽, ▼ Eq. (31); with PDMS (□, △, ▽) or EPR (■, ▲, ▼) as the continuous phase [102]

The above discussion is sufficient to indicate that a promising start has now been made in the understanding of the transport properties of composite polymer membranes when the microscopic component domains are well defined and do not interact appreaciably with one another or with the penetrant.

These conclusions also provide useful additional criteria for the presence of interaction between the component phases. Some instances of this kind have already been noted. Other examples are provided by alternating block PDMS-PC (bis phenol A polycarbonate) copolymers [104], which deviate from the general pattern of Eqs. (30)–(33) increasingly as the PDMS blocks become shorter. Incomplete component phase separation was also indicated by nonconformity to Eq. (27) and other evidence. Analogous complicating factors may come into play in filled polymers [88]. A recent example of this is the unusual diffusion behaviour of CO_2 in membranes consisting of rubber filled with various amounts of carbon black [66].

A simple, even though empirical, alternative treatment in use for the permeability of copolymers and polymer blends is [93]

$$\ln P = v_A \ln P_A + v_B \ln P_B$$

Recently, some physical justification has been offered for this formula as a limiting form of more general mixing rules [105]. The emphasis however, is on application to homogeneous random copolymers and miscible blends.

An approach which is simple and physically informative, though not quantitatively rigorous, involves additive combination of the limiting P and S configurations discussed above [89]. This has recently been done in a manner which allows the extent of participation of each component phase in these configurations to be varied [106]:

$$P = f_A v_A P_A + f_B v_B P_B + \frac{[(1 - f_A)v_A + (1 - f_B)v_B]^2}{(1 - f_A)v_A/P_A + (1 - f_B)v_B/P_B} \tag{34}$$

In Eq. (34), f_A, f_B express in quantitative terms the concept (already referred to above) of the degree of continuity of the respective component phases. Model calculations based on Eq. (34) with various values of f_A, f_B were used to determine the kind of two-phase structure that could reasonably be expected to give rise to the ionic diffusion behaviour observed in cellulose membranes [106] (cf. following section).

6 Membranes with Microscopically Non-Homogeneous Electrical Properties

The chief phenomenon to be considered here is non-uniform distribution of electric charge in charged membranes. The effect of this on ionic sorption and transport properties is of considerable practical interest, because membrane permselectivity for the counterion against the coion (or for uncharged species vs electrolytes), and hence membrane performance in important technical applications (such as electrodialysis) is directly involved.

The membrane systems to be considered include what are normally termed "homogeneous" ion exchangers. Nevertheless, according to the definitions of the introductory section, they should be placed in class (b); because microregions differing in electrical properties can be distinguished, even though this is often done on a purely theoretical basis, as will be seen below.

The partition coefficient S_i of an ionic species i between a uniformly charged membrane and an aqueous solution may be analyzed into electrical and (apparent) chemical affinity factors λ and K_i respectively [107]

$$S_i = K_i \lambda^{z_i} = C_i/C_{i_S} \tag{35}$$

$$\lambda = \exp(-F\Delta\psi/RT) \tag{36}$$

where z_i is the valence of the ion; $\Delta\psi$ represents the difference in electrical potential between the membrane and external solution phases; and F is the Faraday. From Eq.

(35) for the coion ($i = N$) and counterion ($i = P$) respectively we obtain

$$S_P^{z_N} S_N^{z_P} = K_P^{z_N} K_N^{z_P} = K_{PN}^{z_P + z_N} \tag{37}$$

where ideally $K_{PN} = \text{const.}$

The above "uniform charge" or "equipotential volume" model (EVM) treatment assumes a particularly simple form for the sorption of simple ions by highly water-swollen charged membranes. The simple ions in the membrane phase together with the imbibed water are considered to constitute an "internal solution" similar in all respects (except for the electrical potential) to the external one. Then, treating the polymer matrix as essentially inert, $K_{PN} = w$, where w is the water regain or degree of swelling of the membrane.

For a uni-univalent electrolyte in nearly neutral solution ($C_{PS} \simeq C_{NS} = C_S$) and a membrane containing a concentration C_F of univalent fixed charged groups, application of the condition of electroneutrality ($C_P = C_F + C_N$) to Eq. (37) yields

$$S_N^2 + (C_F/C_S)S_N = K_{PN}^2 \tag{38}$$

or

$$S_N = \frac{C_N}{C_S} \simeq \frac{C_N}{C_{PS}} = K_{PN}\left(\left[1 + \frac{\alpha^2}{4}\right] - \frac{\alpha}{2}\right) \tag{38 a}$$

$$\xrightarrow[\alpha \gg 1]{} \frac{K_{PN}}{\alpha} = \frac{K_{PN}^2 C_{PS}}{C_F} \tag{39}$$

where

$$\alpha = C_F/C_{PS}K_{PN} \tag{40}$$

Well known membrane models, such as the TMS model and variants thereof [108], include the additional idealisation that $C_F = \text{const.}$

Real membranes may be expected to deviate from the EVM treatment in two ways: (i) Since the membrane charge arises from fixed ionised groups (point charges), EVM conditions can only be approached when the typical distance between charges is much less than the screening distance or Debye length $1/\sqrt{\beta C_S}$, where $\beta = 8\pi F^2/\varepsilon_D RT$ (ε_D = dielectric constant of the medium). Hence, the EVM will tend to break down at $\alpha \ll 1$. (ii) If the charged groups are not distributed uniformly in the membrane, the EVM will tend to break down even at $\alpha \gtrsim 1$.

In both cases the deviations from the EVM predictions are in the direction of too high S_N values ("incomplete coion exclusion"). In terms of Eqs. (37) and (38) these deviations become evident in practice at low C_S (high α) [109]. This is so, incidentally, only because Eqs. (37)–(38a) are unsuitable for putting into evidence the corresponding discrepancies at high C_S (low α), as has been emphasized recently [110]. Even so, a considerable number of other factors can and has been invoked to account for departures from Eqs. (37) and (38) without abandoning the EVM, including limitations in experimental technique [109] and "binding" or "condensation" of counterions on the charged groups [108]. The latter effect leads to lower S_P or C_F, thus compensating for the higher S_N in Eqs. (37) or (38) respectively. As a result of these alternative interpretations, the EVM is still in wide use for the treatment of ionic sorption properties without proper realization of its severe limitations; in contrast to the interpretation of ionic diffusion properties, where the EVM is hardly ever really adhered to [111]. Attempts to eliminate this dichotomy have, so far, been few [106,112,113].

Accurate modelling of non-equipotential volume behaviour at high C_S is difficult. Accordingly, an idealized equipotential surface model (ESM) is in use. In the simplest version of the ESM (which has the virtue of reducing the problem to its analytically tractable one-dimensional form) first applied by Schofield and Talibuddin [114], the internal solution is considered to be confined within slit-shaped pores of width 2r (the polymer matrix being impenetrable to the sorbed ions). The membrane charge is assumed to be smeared uniformly along the pore walls, where a diffuse electrical double layer following Gouy is set up. The assumption of $K_N = K_{PN} = w$ is also maintained. Then, at sufficiently high C_S and provided that α is also sufficient high, a simple expression for S_N is obtained, namely

$$S_N = w - \frac{2w}{r\sqrt{\beta C_S}} + \frac{4w^2}{\beta r^2 C_E} \qquad (41)$$

In an analogous, though not rigorous, approach [115], the internal solution was considered to consist of an equipotential "Donnan phase" (denoted by subscript or superscript D) next to the pore walls and a "bulk phase" (subscript or superscript B) electrically identical with the external solution (i.e. with $\Delta\psi = 0$) in the centre of the pore. Then, bearing in mind that $S_N^B = w_B = w - w_D$,

$$S_N = w - w_D + (w_D/w)S_N^D \qquad (42)$$

The thickness of the Donnan phase layer r_D was equated to the Debye length; so that $w_D/w = r_D/r = 1/r\sqrt{\beta C_S}$. However, experimental values of w_D/w calculated from measurements of S_P and S_N in cellulose-KCl by means of a combination of Eq. (42), the corresponding equation for the counterion and the proper version of Eq. (37), namely $S_P^D S_N^D = w_D^2$, did not agree very well with the aforesaid theoretical r_D values [116]. Recently, this difficulty has been resolved by showing [110] that Eqs. (41) and (42) become identical under the proper conditions (namely at high C_S), upon setting

$$w_D/w = r_D/r = 2/r\sqrt{\beta C_S} \qquad (43)$$

It was further shown [110] that Eq. (42) in combination with Eq. (43) is, in fact, equivalent to a more general version of Eq. (41), not subject to the restriction of high α, but only to the condition of high C_S. The latter condition is expressed more accurately in the form $r \geq 2r_D$. At lower C_S, i.e. higher r_D, values the formulation of Eq. (42) may still be employed, but w_D becomes a complicated implicit function of both C_S and α [110].

Among recent ionic partition data, those referring to weakly charged membranes, like cellulose [110,117], are particularly useful; because it is difficult to claim validity of the EVM at high C_S (cf. mean estimated distances of separation between fixed $-CO_2^-$ groups in viscose sheet [110] well in excess of a Debye length of ~ 1 nm at $C_S \sim 0.1$ mol dm^{-3}). Accordingly, the conclusion of Ref. [117], that the observed behaviour of S_{Cl} conforms to the EVM at higher C_S must be attributed to the use of Eq. (38); which, like Eq. (37), is unsuitable for showing up deviations from the EVM, as already pointed out above. In Ref. [110], on the other hand, data on both S_{Na} and S_{Cl} for cellulose membranes were treated on the basis of Eqs. (42), (43) in the manner already indicated above in connection with Ref. [116] taking care to allow also for some deviation from the model assumption $K_{PN} = w$, due to the presence of nonsolvent imbibed water (i.e. water of hydration of the polymer chains and charged groups) or other causes. The experimental w_D values determined in this manner exhibited the trend required by Eq. (43), which was not materially altered by the precise choice of K_{PN}. There was also reasonable agreement with the corresponding data of Ref. [116]. At the quantitative level, the said w_D values tended to be too low and to show some dependence on C_F (which was varied by introducing varying amounts of $-SO_3^-$ groups though absorption of a suitable anionic dye [110]), contrary to the implications of Eq. (43) (cf. Fig. 13). This is not difficult to understand, however, in view of the high degree of idealization introduced by the equipotential surface principle [110]. Thus, it appears that the ionic sorption properties at high C_S can be interpreted semiquantitatively in a reasonable manner in terms of the ESM. Alternative interpre-

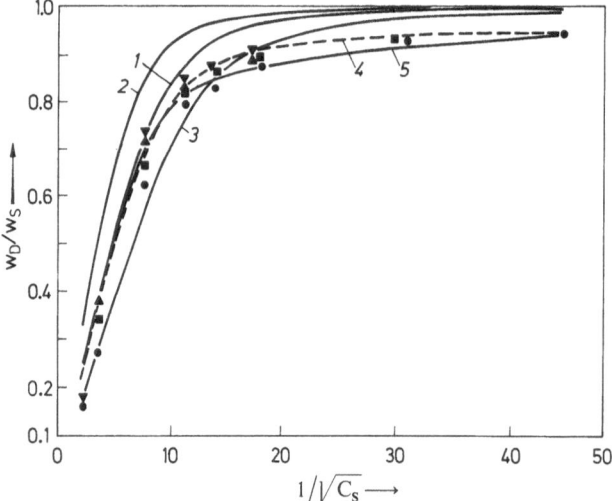

Fig. 13. Experimental values of w_D/w_s ($K_{PN} = w_s = 0.93$ w, assuming 7% of "nonsolvent" imbibed water), calculated as described in the text, for the sorption of NaCl at pH $\simeq 5.8$ and 22 °C by cellulose membranes undyed (●) or dyed with C.I. Direct Blue 1 (■, ▲, ▼): $C_F/w = 0.025$ (●), 0.044 (■), 0.069 (▲), 0.085 (▼) mol/dm^3 (when all fixed carboxyl and sulphonic acid groups are fully ionised). Lines were calculated by the ESM for slit-shaped pores (see text) for the same values of C_S and C_F/w assuming (i) charged pores of radius r = 6 (line 1), 4.5 (line 2), or 8 (line 3) nm; (ii) charged pores of r = 6 nm (95% of w_s) + uncharged pores (5% of w_s) (line 4); (iii) narrow + wide pores of equal surface charge density with 5.25 nm (86.5% of w_s) and 40 nm (13.5% of w_s) (line 5) [110]

tations which preserve the EVM were examined in detail, but were found to be less satisfactory and less realistic physically [110].

The tendency towards deviation from the EVM would be expected to be enhanced in more hydrophobic membranes, because of the lower dielectric constant and hence shorter Debye length in such media. Polyamide membranes in contact with acidified external solutions are of considerable interest in this respect, in view of the fact that the EVM treatment based on Eqs. (35)–(37) has been used with considerable success [107] as an alternative way of describing the uptake of acid dyes by polyamides. (For the more common way of treating these systems see the previous section but one). This treatment implies that the electrical interactions in ionic sorption will be governed by the net charge of the membrane, which for polyamides is positive at pH values below the isoelectric point (i.e. below pH $\sim$ 6). Recent partition measurements of Na^+ and Cl^-, however, show [118] that both S_{Na} and S_{Cl} tend to increase with diminishing C_S even under conditions where there is a marked preponderance of $-NH_3^+$ over $-CO_2^-$ groups. The behaviour of S_{Na} is obviously the exact opposite of "coion exclusion", thus indicating breakdown of the EVM even at the qualitative level.

Deviations from EVM behaviour at low C_S (high α) values were first described successfully in terms of a non-uniform charge distribution by Glueckauf [112], who postulated the existence of equipotential microregions each characterized by a local concentration of fixed charged groups C_{FL}. It was further postulated that C_{FL} conforms to the following distribution function

$$f(C_{FL}) \, dC_{FL} = k_0 C_{FL}^{-q} \, dC_{FL} = \frac{1-q}{B^{1-q} - A^{1-q}} \, C_{FL}^{-q} \, dC_{FL} \tag{44}$$

where $A \leq C_{FL} \leq B$ is, for greater precision and generality, treated as the local concentration of counterions in the absence of any sorbed electrolyte; $0 \leq q \leq 1$ is a constant which provides a measure of the extent of non-homogeneity of the charge distribution ($q = 0$ corresponding to uniform charge); A and B are treated as constant and k_0 is a normalisation constant. Eq. (44) leads to the following expressions for the overall values of C_F and S_N which are observed in practice:

$$C_F = k_0 \int_A^B C_{FL}^{1-q} \, dC_{FL} = \frac{(1-q)(B^{2-q} - A^{2-q})}{(2-q)(B^{1-q} - A^{1-q})} \tag{45}$$

$$S_N = k_0 \int_A^B S_{NL} C_{FL}^{-q} \, dC_{FL} = \frac{(1-q) I(\alpha_A, \alpha_B)}{B^{1-q} - A^{1-q}} C_S^{1-q} \tag{46}$$

where the local value S_{NL} of the coion partition coefficient is evaluated in terms of $\alpha_L = C_{FL}/K_{PN}C_S$ through Eq. (38a), which is valid locally (i.e. within each microregion); and $\alpha_A = A/K_{PN}C_S$, $\alpha_B = B/K_{PN}C_S$. As shown in Ref. [112], $I(\alpha_A, \alpha_B)$ is very nearly independent of C_S over quite an extensive range in the $\alpha \gg 1$ region. Thus, the dependence of S_N on C_S in this range is described by a power law. At still higher α, there is a transition to simple proportionality, since Eq. (39) must be valid at sufficiently high α.

There is no a priori physical basis for Eq. (44). In fact, the distribution function of C_{FL} was induced directly from the power law of Eq. (46); which was, in turn, established empirically on the basis of coion sorption measurements on certain ion exchange membranes [119]. A wide variety of ion exchange membranes have, since that time, been shown to conform to this power law behaviour. Values of q extending over practically the full allowed range have been recorded [120]. The highest values of q indeed correspond to ion exchangers known to be highly heterogeneous (cf. Table 4). However, a precise correlation between q and specific structural or other membrane characteristics is still lacking. Among ion exchange membranes studied in this manner, a series of copolymers of polystyrene sulphonic acid grafted onto a polyethylene matrix are noteworthy [120]. In this case, the values of q determined from sorption of SO_4^{--} by the H^+ form of the membranes were confirmed by autoradiography on the radioactively labelled Cs^+ form (cf. Table 4). The ratio C_F/B was determined from the ratio of the maximum and average photometric readings of the autoradiographs respectively. Eq. (45) was then applied assuming $B \gg A$. It is, however, regrettable that no attempt was apparently made to validate Eq. (44) by extracting information about the detailed C_{FL} distribution function from these photometric readings.

Measurements of the sorption of Cl^- by regenerated cellulose membranes [110,121] can serve as examples of the applicability of the Glueckauf model to weak ion exchangers. Furthermore, the hydrophilic polymer matrix of these membranes might be expected to lead to more homogeneous sorption properties. The observed values of q are indeed relatively low (cf. Table 4) by comparison with most common ion exchangers. (The fact that q is not still lower can be attributed to morphological heterogeneities in commercial regenerated cellulose membranes [121]). The work reported in Ref. [121] should call attention to the importance of monitoring the value of C_F, which is considered to be independent of C_S in the Glueckauf treatment {cf. Eq. (45)}. This assumption was taken for granted in all previous experimental work. In Ref. [121] it is pointed out that, if the variation of C_F with C_S is appreciable but conforms to the empirical relation $C_F = KC_S^n$ (K, n = constants), then the power law of Eq. (46) is maintained. However, the resulting value of q, requires correction, the corrected value being [121]

$$q_0 = (q - n)/(1 - n) \tag{47}$$

As indicated in Table 4, both the q and q_0 values determined for cellulose membranes in Ref. [121] were remarkably constant irrespective of the presence or absence of added $-SO_3^-$ groups.

The main questions which arise with respect to the Glueckauf model discussed above concern the uniqueness and the physical justification of the postulated local charge variation represented by Eq. (44). With reference to the former question, some model calculations of S_N for $f(C_{FL})$ distributions of different shapes and widths may be noted [122,123]. It was shown that the magnitude and form of the deviation from EVM behaviour for a given B/A ratio is determined largely by the shape of the $f(C_{FL})$ function near the lower limit A. The precise form of $f(C_{FL})$ in the higher C_{FL} region appears to have a relatively minor effect on the conformity of S_N to the power law of Eq. (46) or the value of q.

Table 4. Recent interpretations of coion sorption and diffusion based on the Glueckauf model [110,120]

Membrane	Charged Groups	C_F/w	q		q_0	$D \times 10^7$	$\dfrac{D}{x} \times 10^7$		Coion
		mol/dm³	(a)	(b)		cm²/s	cm²/s (a)	(b)	
PE-PSSA 2% DVB	$-SO_3^-$	1.97	0.60	0.53		7.6	0.45	2.4	SO_4^{--}
5% DVB	$-SO_3^-$	2.96	0.58	0.52		5.2	0.47	3.8	
10% DVB	$-SO_3^-$	3.94	0.57	0.50		3.2	0.58	5.2	
15% DVB	$-SO_3^-$	4.21	0.64	0.59		3.6	0.75	4.4	
2% DVB	$-SO_3^-$	0.57	0.78			15.3	8.1	4.8	
Ionac MC-3470	$-SO_3^-$	3.36	0.91			25.1	4.9	1.6	
Cellulose (undyed)	$-CO_2^-$	0.025	0.42		0.34				Cl^-
+0.75% Dye A	$-CO_2^-$, $-SO_3^-$	0.044	0.43		0.31				
+0.66% Dye A	$-CO_2^-$, $-SO_3^-$	0.069	0.42		0.33				
+2.35% Dye A	$-CO_2^-$, $-SO_3^-$	0.085	0.39		0.30				
+2.15% Dye B	$-CO_2^-$, $-SO_3^-$	0.068	0.43		0.33				

Notes: Col. 1. – (i) PE-PSSA: polystyrene sulphonic acid grafted onto a polyethylene matrix and containing indicated amounts of divinyl benzene [120]; (ii) Ionac MC-3470 (Ionac Corp., USA): powdered mixture of PSSA ion exchanger and inert polymer compressed hot with embedded mesh of supporting material [120]; (iii) Cellophane PUT 600/23 sheet (British Cellophane Ltd.) washed to remove plasticizer [110]; (iv) Dye A : C. I. Direct Blue 1 (tetrasulphonate substantive dye)[10]; (v) Dye B : Procion Yellow HA (disulphonate reactive dye, I.C.I. Ltd.) [121].

Col. 3. – Values given apply when all fixed acid groups are fully ionised.

Col. 4, 5. – q from (a) Eq. (46) (b) autoradiography (see text).

Col. 6. – q_0 from Eqs. (46) and (47).

Col. 7–9. – Mean diffusion coefficients of H_2SO_4 in concentration range $C_S = 0-0.1$ mol dm^{-3}. x obtained from (a) Eq. (48) or (b) Eq. (48) with v_p replaced by v_G given by Eq. (49)

Concerning the second question raised above, a general physical justification of Eq. (44) was offered in Glueckauf's paper [112] in terms of "aqueous fissures" (actually observed in his membranes by electron microscopy), of which [112] "there would be a whole range of widths, the narrow ones being much more numerous than the wider ones, so as to conform to the observed counterion distribution pattern of Eq. (1.6)". {Eq. (1.6) in Ref. [112] is equivalent to Eq. (44) here}. The ESM offers obvious possibilities of checking this general statement in a concrete, physically realistic manner. Very recently [121], model calculations of S_N based on the general ESM equations have been reported both for single pores and for assemblies of pores of two different widths. In the latter case, and more specifically for pore assemblies involving a small proportion of wide pores, conformity of S_N to the power law of Eq. (46) was found over quite extensive ranges of C_S (cf. Fig. 14). No such conformity to Eq. (46) at any C_S was detected in the case of single pores. These results throw considerable light both on the physical significance of the Glueckauf treatment and on the usefulness of the ESM. Calculations involving assemblies of pores with different surface charge densities have also been reported [121].

Direct application of ESM calculations of this kind to the interpretation of experimental coion behaviour has been reported [110,121]. The need for assuming the presence of an appreciable proportion of either very wide or uncharged pores to interpret the ion sorption data of Ref. [110] at low C_S is illustrated in Fig. 13. The latter interpretation was also advanced for the analogous Cl^- sorption data of Ref. [117]. (Incidentally, the fraction of imbibed water thus estimated to be in regions of practically zero electric charge amounts to $\sim 5\%$ of w; and so considerably exceeds what could

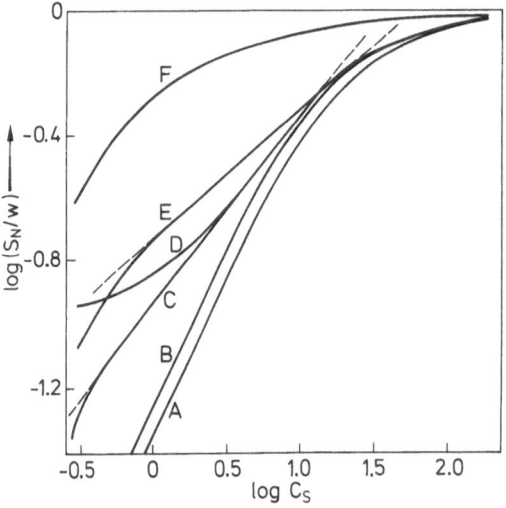

Fig. 14. Coion partition coefficients S_N for uni-univalent electrolyte, $C_F/w = 21.7$ mmol/dm³, $K_{PN} = w$, calculated by the EVM (line A) or the ESM for slit-shaped pores (lines B-F): (i) Single pores of radius r = 6 (line B) or 40 (line F) nm. (ii) Assemblies of two pores of equal surface charge density of mean radius 6 nm and individual radii (in nm) 5.306, 40 (line C); 5.454, 80 (line D); 4.583, 40 (line E); where 98% (C), 99% (D), or 96% (E) of the total pore wall surface area is in narrow pores [121]. Note that the power law of Eq. (46) is obeyed over a considerable range in cases C and E

plausibly be attributed to experimental error due to films of external solution adhering to the membrane [110]).

The treatment of ionic diffusion properties is a more difficult task. The obstruction effect of the polymer matrix can be represented by a number of semiquantitative equations, the most popular of which is [111]

$$\varkappa_i = D_i/D_i^0 = (1 - v_P)^2/(1 + v_P)^2 \tag{48}$$

where D_i, D_i^0 represent the diffusion coefficient of ionic species i in the membrane and in aqueous solution respectively and v_P is the volume fraction of the polymer matrix.

However, the question of main interest is the relation between D_P and D_N (or P_P and P_N) and their dependence on C_S. Glueckauf et al. [112,124] treated P_N of a number of ion exchange membranes by means of the Bruggeman equation generalized to a large number of component phases {Eq. (31) is the two-component version}, the composition being given by Eq. (44). The experimental P_N for the membranes examined differed both in absolute value and in their detailed functional dependence on C_S. The latter aspect of the data could be interpreted on the basis indicated above by assuming that the continuous component phase (cf. previous section) in each membrane was different. The results obtained in this way were consistent with morphological evidence furnished by electron microscopy [124]. Thus the membrane assigned the low C_{FL} microregions as continuous phase was found to possess a network of relatively wide water-filled microfissues; whereas the membrane, in which the high C_{FL} regions were identified as the continuous phase, exhibited water-filled cavities or pockets isolated from one another. These structural features also accounted for the much lower P_N values in the latter case {after allowing for the effect of different water uptake according to Eq. (48)}, even though the corresponding values of S_N were comparable. A structure involving a continuous high C_{FL} phase is believed to exist in the newer, technically important perfluorosulphonate ("Nafion") cation exchangers. There is considerable evidence [125] that, in these membranes, the internal solution is in the form of clusters (most probably spherical in shape [126]). It is further suggested [126] that, when neighbouring clusters are sufficiently close to one another, they may be interconnected by means of narrow channels, as illustrated in Fig. 15; thus giving rise to the aforementioned continuous high C_{FL} component phase. The dependence of the ion transport properties of the membrane on the amount of imbibed water can be interpreted semiquantitatively by assuming random location of the clusters and applying percolation theory [126]. Detailed studies of ionic permeabilities or diffusion coefficients as a function of C_S are lacking, however, and no link-up with the aforementioned work of Glueckauf et al. was made.

A study of coion diffusion (mean diffusion coefficient of H_2SO_4 over the range $C_S = 0 - 0.1$ mol dm^{-3}) in a series of cation exchange membranes of different imbibed water content has been reported [113]. On the basis of Eq. (48) one expects $\bar{D}_i/\varkappa_i = D_i^0 = \text{const}$. As illustrated by the examples included in Table 4, this expectation is borne out semiquantitatively, except for the strongly heterogeneous (characterized by large q values) membranes (Table 4, col. 8). However, this deviation was largely corrected (Table 4, col. 9) by considering that the sorbed electrolyte is excluded not only from the polymer matrix, but also from the high C_{FL} part of the

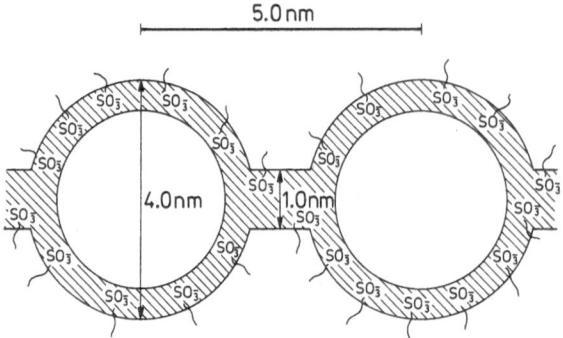

Fig. 15. Cluster network model for highly cation-permselective "Nafion" membranes [126]. Counterions are largely concentrated in the high-charge shaded regions which provide somewhat tortuous, but continuous (low activation energy), diffusion pathways. Coions are largely confined to the central cluster regions and must, therefore, overcome a high electrical barrier, in order to diffuse from one cluster to the next

internal solution. The latter was defined by $C_{FL}/w > C_S$ and was assumed to occupy a fraction v_G of the total membrane volume, which was calculated by means of the Glueckauf model, namely by

$$v_G = (1 - v_P) \{1 - k_0 C_F^{1-q}/(1 - q)\} \tag{49}$$

Then assuming for simplicity that $C_N = 0$ within v_G and $C_N = C_S$ elsewhere in the internal solution, v_P in Eq. (48) (and in other analogous equations [113]) was replaced by $v_P + v_G$.

Diffusion coefficients of Na^+ (counterion) and Cl^- (coion) in cellulose membranes have recently been determined by permeation measurements either under an electrolyte concentration gradient [117] or by means of tracers in the absence of such a gradient [106]. In both cases, it was found that, at low α, $D_{Cl}/D_{Na} = 1.54$, in close agreement with the corresponding mobility ratio in aqueous solution. However, both D_{Na} and D_{Cl} vary with C_S much more markedly than in aqueous solution. D_{Na} shows typical counterion behaviour [111], i.e. decreases with diminishing C_S. There is no analogous typical pattern of behaviour for coions [111]. D_{Cl} shows a trend largely paralleling, but less marked than, that of D_{Na} [106,117]. The data of Ref [106] further show a tendency for reversal of this trend at sufficiently low C_S.

To interpret the above behaviour of D_{Na} and D_{Cl} qualitatively or semiquantitatively in a manner consistent with sorption behaviour, a model was constructed on the basis of the division of the internal solution into "Donnan" and "bulk" phases [106]. These component phases were further assumed to be arranged partly in series and partly in parallel, in accordance with Eq. (34) (phases A and B therein being identified with phases D and B here). This approach offers a somewhat crude but simple and flexible way of representing the two kinds of diffusion pathway inherent in previous qualitative ideas [111] and in other models [112,125] (cf. also Fig. 15), namely paths of sensibly constant electrical potential (corresponding to the parallel configuration here) and paths requiring passage across electrical potential barriers (corresponding to the serial configuration here). On the basis of this model, it was shown [106] that the latter type of pathway leads to decrease of D_i with diminishing C_S; the effect being more pronounced for the counterion, if the "bulk" phase has a greater "degree of continuity" f_B (cf. previous section). It was also pointed out that the tendency of D_{Cl} to increase at high dilution can be understood in terms of a network made up of the unusually wide and/or uncharged pores already invoked for the interpretation of ionic sorption behaviour at low C_S. This simple model highlights the important point that the generally observed [111] decrease of D_P with diminishing C_S can be interpreted as an effect of electrical potential nonhomogeneity; and is, therefore, not necessarily indicative of cation binding or condensation on the fixed charged groups, as has been believed hitherto [108]. An analogous model involving parallel and serial combinations of (slit-shaped) pores characterized by different diameters and surface charge densities was also formulated in conjunction with the ESM [106].

It is important to point out that sorption of simple ions is not affected only by charge non-uniformity. The non-homogeneity of the dielectric constant in the membrane is also significant. This becomes particularly evident in the case of uncharged hydrophobic membranes of the type commonly used for water desalination by reverse osmosis, where "dielectric exclusion" of the electrolyte from the membrane phase is the only sorption mechanism in operation [127]. If the swollen membrane is treated as a homogeneous phase, the ionic partition coefficient S_i is expected to be

$$ S_i = \exp\left[-\frac{z_i F^2}{2 b_i RT} \left(\frac{1}{\varepsilon_{DM}} - \frac{1}{\varepsilon_{DB}} \right) \right] \tag{50} $$

where b_i is the ionic radius and ε_{DM}, ε_{DB} are the dielectric constants of the membrane and bulk solution phases respectively. The S_i values calculated by Eq. (50) usually turn out to be considerably lower than the observed ones [128]. This is accounted for by considering the imbibed water to be at least partly in the form of clusters or pores in which the sorbed ions are preferentially located.

An approximate quantitative treatment was first proposed by Glueckauf [127] in terms of a cylindrical pore model which yields

$$ S_i = \exp\left[-\frac{z_i F^2 \gamma (1 - \varepsilon_{DA}/\varepsilon_{DB})}{2\varepsilon_{DB} RT \{ r + (1 - \gamma) b_i (1 - \varepsilon_{DA}/\varepsilon_{DB}) \}} \right] \tag{51} $$

where $\gamma = (1 + \varkappa^2 r^2)^{-1/2}$; r is the radius of the pore; $\varkappa$ is the Debye length in aqueous environment; and ε_{DA} is the dielectric constant of the polymer matrix of the membrane (assumed impenetrable to ions as in previous models). The reader may also consult a review on this subject [129]. Here, we note some measurements of S_i for NaCl, KCl and $MgCl_2$ in cellulose acetate (of 39.8 % acetyl content) membranes for the purpose of testing Eq. (51) [130]. It was concluded that the values of S_i (considering the cation to be the controlling ion) and the dependence of S_{NaCl} on C_S could be interpreted with the aid of Eq. (51), to a first approximation, by assuming $r \sim 0.17$–0.24 nm [130]. This conclusion is subject to the qualification that cellulose acetate membranes ordinarily also possess non-negligible electric charge [131]. In any case the above values of r cannot be taken literally in view of the highly idealised and approximate nature of the model. The same may be said of the pore radius values deduced by comparison of hydraulic permeability and tritiated water diffusivity of cellulose acetate membranes [132,133]. Even so, it is remarkable that these widely different methods of estimating r should give results of the same order of magnitude.

The effect of the scale of dielectric non-homogeneity has been studied theoretically in some detail for a swollen membrane of $\varepsilon_{DA} = 7.09$ in which 50% of the volume is occupied by imbibed water ($\varepsilon_{DB} = 78$) [134]. This was done on the basis of a serial type model in which a test ion is surrounded by alternating spherical shells of phases A and B of thickness h. S_i was found to increase quite markedly as h was varied from $h \ll b_i$ ($\sim$ homogeneous medium) to $h \gg b_i$ ($\sim$ macroscopically non-homogeneous medium). Most of the change in S_i was observed to occur in the region of $h \gtrsim b_i$.

The discussion given in this section shows that non-homogeneity of membrane electrical properties is widespread and markedly influences ionic sorption and diffusion behaviour. Proper understanding of these effects is, therefore, important and may be expected to contribute materially to the design of more highly permselective membranes.

7 Macroscopically Non-Homogeneous Membranes

Membranes classified under (c) in the introductory section are of considerable importance. Most natural membranes are macroscopically non-homogeneous. In artificial membranes non-homogeneity may be introduced either deliberately (laminates, "asymmetric" membranes) or spuriously (e.g. "skin layers" on films made by extrusion). Variation of S and D_T across the membrane, i.e. in the X direction, is of particular interest; non-homogeneity along the plane of the membrane is important in certain special cases, e.g. charged mosaic membranes, which are not of immediate interest here. Also asymmetric membranes prepared for the sole purpose of producing an ultrathin "active layer" to maximize permeation flux are outside the scope of the present discussion.

7.1 Steady State Permeability Properties

In the absense of any dependence of S and D_T on penetrant concentration, the sorption and steady-state permeation properties of the membrane are described by effective solubility and permeability or diffusion coefficients given by

$$S_e = \frac{1}{1} \int_0^1 S(X)\, dX \tag{52}$$

$$P_e = 1 \Big/ \int_0^1 dX/P(X) = S_e D_e \tag{53}$$

where l is the thickness of the membrane. For a two-component laminated medium, Eqs. (52) and (53) reduce to Eqs. (27) and (29) respectively, which can easily be generalised to any number of components.

The situation, is more complex if S and/or D_T are also functions of C or a. Eq. (52) is still valid; but, if P(a, X) is not separable into a product of functions of a and X, P_e cannot, in general, be expressed analytically. Under these conditions, the value of P_e will, in general, also differ according to whether flow is in the $+X$ or $-X$ direction. This directional permeability property has attracted considerable interest, as indicated in a previous review [135].

Further developments since that time include theoretical studies [136, 137] to establish the conditions for maximization of the directional permeability or "flow asymmetry" effect. This work demonstrated in a simple manner that high flow asymmetry factors $f_a = \vec{P}_e / \vec{P}_e$ can, in principle, be achieved under suitable conditions in contradistinction to what was implied by earlier studies [135]. Furthermore, previous experimental data on asymmetric membranes were interpreted semiquantitatively by means of suitable model calculations [137]. Thus, it was shown that the value of $f_a = 3.4$ obtained for water vapour permeating through a laminate of Nylon 6-ethyl cellulose (Etho-cell 610) [138] was reasonable and close to the maximum achievable with those polymeric materials. On the other hand, a value of $f_a \simeq 6$ (or even more) measured at higher upstream water vapor boundary pressures in a polyethylene membrane with a gradation in vinyl alcohol graft content [139], could not be accounted for, except by assuming a mechanism whereby micro-cracks are generated in the membrane by the strong differential swelling stresses set up when the high water vapour pressure is on the side rich in the graft component. (Microcrack opening by this mechanism has now been put in evidence in the case of water permeating through polypropylene grafted with acrylic acid) [140]

7.2 Permeation Time Lags

It is well known that insertion of the above effective coefficients S_e and P_e or $D_e = P_e/S_e$ into Eqs. (2) or (3) respectively, does not lead to the correct description of transient diffusion. However, the behaviour of the ideal Fickian system defined by S_e and P_e or D_e constitutes a useful standard of reference. Given the appropriate theoretical background, one may then deduce information about S(X), $D_T(X)$ from the nature and magnitude of the deviation of suitable observed kinetic parameters from the calculated Fickian values.

This kind of treatment was first applied to permeation time lags. Development of the theoretical background has already been reviewed [135], including experimental verification of time lag expressions for laminated media of the AB type. Laminated media are particularly useful for this purpose, because the full S(X), $D_T(X)$ functions can be determined accurately by measuring the solubility and diffusion coefficients in each component of the laminate separately. Some further data of this type may be noted [141]. However, the principal development to be reported here concerns the experimental application of the full method of time lag analysis to gases permeating through porous membranes presumed to possess macroscopically non-homogeneous structure [142–145]

Typical data for the application of this method are shown in Fig. 16. They refer to standard permeation experiments in which the upstream (at X = 0) and downstream (at X = 1) surfaces of the membrane are maintained at constant penetrant activities

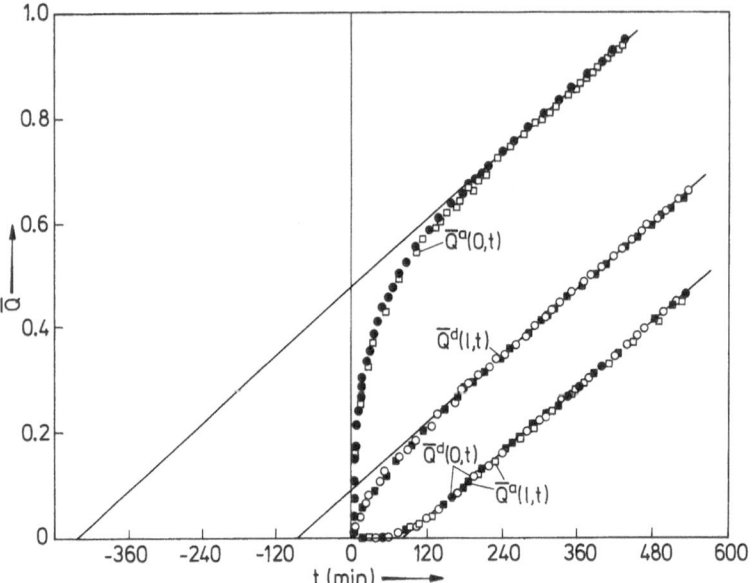

Fig. 16. Experimental permeation curves for N_2 at 297 K in Membrane D of Table 5 (points) and their corresponding steady-state asymptotes (lines) with Q given in dimensionless from $\overline{Q}$ [145]. Absorptive permeation: ● forward (+X) flow; □ reverse (−X) flow. Desorptive permeation: ○ forward flow; ■ reverse flow. Note conformity of the data to Eqs. (68)

a_0 and a_1 respectively. One measures the amount of penetrant, $Q(0, t)$, entering the membrane at $X = 0$, as well as the amount, $Q(1, t)$, which leaves the membrane at $X = 1$. The membrane is pre-equilibrated at uniform penetrant activity a_1 ("absorptive permeation" denoted by superscript a) or a_0 ("desorptive permeation", superscript d). Thus, it is possible to determine two "upper" permeation curves $Q^a(0, t)$, $Q^d(1, t)$ and two "lower" permeation curves $Q^a(1, t)$, $Q^d(0, t)$, as shown in Fig. 16. The steady-state linear asymptotes (denoted by Q_s) yield four time lags (intercepts on the t axis; cf. Fig. 16) $L^a(0) < 0$, $L^d(1) < 0$ and $L^a(1) > 0$, $L^d(0) > 0$ respectively. The same measurements may be repeated for flow in the −X direction (denoted by Q^*, L^*).

In the experiments described here $a_1 = 0$ and the expressions for P_e, D_e and S_e are [144, 145]

$$dQ_s(0, t)/dt = dQ_s(1, t)/dt = UP_e a_0/l \qquad (54)$$

$$L^a(1) - L^a(0) - L^d(1) + L^d(0) = l^2/D_e \qquad (55)$$

$$S_e = P_e/D_e \qquad (56)$$

where U is the cross-sectional area of the membrane.

It is noteworthy that the effective coefficients P_e, D_e and S_e are obtainable in this way without recourse to equilibrium sorption measurements. This amounts to a kinetic method of measuring S_e, exactly analogous to that applicable to ideal systems (where measurement of $L^a(1)$ suffices for this purpose). This method is also applicable to concentration dependent systems, as has been shown both theoretically and experimentally [1,4,147-149].

Consideration has also been given to proper experimental design for the measurement of $Q^d(l, t)$ [147-148].

The time lags defined above are, for convenience examined in the form: [1,4,143-145]
$L^a \equiv L^a(l)$, $\Delta L^a \equiv L^a - L^a(0)$, $\delta L \equiv L^a - L^d(l)$ and $\delta \Delta L \equiv \Delta L^a - L^d(l) + L^d(0)$.
The corresponding "Fickian" parameters calculated on the basis of S_e and P_e or D_e are designated L_s^a, ΔL_s^a, δL_s and $\delta \Delta L_s$ respectively; and, for concentration independent systems (like the ones considered here), are given by

$$6L_s^a = 2\Delta L_s^a = 2\delta L_s = \delta \Delta L_s = \delta \Delta L = l^2/D_e$$

When analysing the discrepancies (L_E^a, etc.) between the observed time lag parameters (L^a, etc.) and the calculated ideal values (L_s^a, etc.), one must take into account possible contributions from causes other than dependence of S, D_T on X. Analysis of the case of simultaneous dependence of S, D_T on X and t shows that [4,135]

$$L_E^a = L^a - L_s^a = L_T^a + L_h^a$$

where L_E^a is the observed discrepancy or "non-Fickian time lag increment" and L_T^a, L_h^a are the corresponding increments due to time- and X-dependence respectively. Exactly analogous expressions hold for ΔL_E^a, δL_E ($\delta \Delta L_E = 0$). Study of the properties of the individual time-lag increments shows [1,4,135] that $\Delta L_T^a = 0$, and for S, D_T independent of penetrant concentration

$$\Delta L_h^a = -\delta L_h = \delta L_h^* = -\Delta L_h^{a*} \tag{57}$$

These properties enable one to separate time dependence from X dependence effects e.g. through the relation [144,145]

$$\delta L_E = \delta L_h + \delta L_T = -\Delta L_h^a + \delta L_T = -\Delta L_E^a + \delta L_T \tag{58}$$

The most detailed and accurate data so far have been obtained with N_2. As illustrated in Table 5, δL_E, ΔL_E^a obey Eq. (57). Therefore, $\delta L_T = 0$ according to Eq. (58). The properties of L_h^a, ΔL_h^a and δL_h have been shown [4,135] to be determined by the function $S(X)^2 D_T(X)$. Thus, $\Delta L_E^a = 0$ for membranes B and E in Table 5, indicates that $S(X)^2 D_T(X)$ is nearly symmetrical about the membrane midplane ($X = 1/2$) [4,135], in accord with the symmetrical mode of construction of these membranes [143,145]; the algebraic sign of L_E^a implies [4,135] diminution of $S(X)^2 D_T(X)$ from the surfaces inward. Similarly, the properties of $\Delta L_E^a/\Delta L_s^a$ indicate that in the remaining asymmetric membranes (A, C, D, F, G) there is a predominant tendency for $S(X)^2 D_T(X)$ to decrease with increasing X; the degree of non-homogeneity appears to be greatest in the case of D and least in the case of C and F [145]. The advantage of using porous membranes is that the spatial variation of S and D_T is relatively simply related to that of the local porosity [146]. The latter exhibited a rather complicated two-dimensional pattern; nevertheless, there was a clear overall spatial variation in the X direction in line with the inferences drawn above from time lag analysis [145]. (Spatial dependence

Table 5. Permeation results for N_2 at 297 K in macroscopically nonhomogeneous porous graphite membranes [143-145]

Membrane	A	B	C	D	E	F	G
Mode of construction	non sym	sym	non sym	non sym	sym	non sym	non sym
Thickness (cm)	1.64	1.62	1.13	1.19_3	0.84_3	0.49_5	0.97_4
Porosity	0.15	0.13	0.13	0.15	0.15	0.15	0.13
$10^5 P_e$ (cm²/s)	7.8_0	8.8_1	2.6_4	2.4_9	4.8_5	4.3_0	3.0_5
S_e { (a)	0.87	0.86	0.86	0.74	0.74	0.82	0.87
$\quad$ (b)	0.91	0.86	0.86	0.72	0.75	0.82	0.84
L_E^a/L_s^a	−0.26	−0.38	−0.18	−0.31	−0.37	−0.09	−0.18
$\Delta L_E^a/\Delta L_s^a$	0.30	−0.05	0.16	0.52	−0.03	0.19	0.26
$\delta L_E/\delta L_s$	−0.27	0.05	−0.16	−0.52	0.03	−0.19	−0.26
L_E^{a*}/L_s^a	−0.26	−0.38	−0.17	0.32			
$\Delta L_E^{a*}/\Delta L_s^a$	−0.27	−0.05	−0.21	−0.51			
$\delta L_E^*/\delta L_s$	0.31	0.05	0.20	0.51			

Notes: (a) S_e obtained from equilibrium sorption measurements;
(b) S_e obtained from permeation measurements by Eqs. (54)-(56)

of S and D_T purely in the radial direction does not lead to deviations from ideal time lag behaviour).

7.3 Transient Diffusion Kinetics

A similar method of analysis of transient state diffusion kinetics has been proposed [144, 149] based on the consideration that, in any experiment, the kinetic behaviour of the system represented by S(X), $D_T(X)$ will generally deviate from that of the corresponding ideal system represented by S_e, D_e in either of two ways: (i) ideal kinetics is obeyed, but with a different effective diffusion coefficient D_n, where n = 1, 2, ... denotes a particular kinetic regime (D_n is usually deduced from a suitable linear kinetic plot); or (ii) ideal kinetics is departed from, in which case one is reduced to comparison between the (non-linear) experimental plot and the corresponding calculated ideal line.

The following effective diffusion coefficients D_n may be defined on the basis of standard sorption and permeation experiments [149-151] (when absorption or desorption conditions need to be specified, superscripts a or d respectively are again used):

(a) *Sorption*
If Q_t, Q_∞ are the amounts of penetrant sorbed at time t and at equilibrium respectively for unsymmetrical sorption [144, 150] (membrane exposed to penetrant at X = 0, blocked at X = l) we have

$$Q_t/Q_\infty = 2(D_1 t/\pi l^2)^{1/2} \tag{59}$$

$$\ln(1 - Q_t/Q_\infty) = I_2 - \pi^2 D_2 t/4l^2 \tag{60}$$

at short and long times respectively. The corresponding expressions for symmetrical sorption (membrane exposed to penetrant at both $X = 0$, $X = l$) are obtained by substituting $1/2$, for l in Eqs. (59) and (60); and the respective effective diffusion coefficient and intercept values are denoted by D_{1M} and D_{2M}, I_{2M}.

(b) *Permeation*

For the lower permeation curves, and noting that $Q^a(l, t) = Q^d(0, t)$ [see Eq. (68) below] we have

$$\ln [\sqrt{t}\ dQ^a(l, t)/dt] = \ln (2USa_0 \sqrt{D_4/\pi}) - l^2/4D_3t \tag{61}$$

$$\ln \{[Q^a(l, t) - Q^a_s(l, t)]/Q_\infty\} = I_5 - \pi^2 D_5 t/l^2 \tag{62}$$

at short and long times respectively.
For the upper permeation curves $Q^a(0, t)$, $Q^d(l, t)$ we have, e.g. for $Q^a(0, t)$

$$Q^a(0, t)/Q_\infty = 2(D_1^a t/\pi l^2)^{1/2} \tag{63}$$

$$\ln \{[Q^a_s(0, t) - Q^a(0, t)]/Q_\infty\} = I_7^a - \pi^2 D_7 t/l^2 \tag{64}$$

at short and long times respectively. Alternatively, use may be made of the net amount sorbed during permeation $\Delta Q_t = |Q(0, t) - Q(l, t)|$ and $\Delta Q_\infty = |Q_s(0, t) - Q_s(l, t)|$ in which case we have

$$\Delta Q_t/\Delta Q_\infty = 4(D_6 t/\pi l^2)^{1/2} \tag{65}$$

$$\ln (1 - \Delta Q_t/\Delta Q_\infty) = I_8 - \pi^2 D_8 t/l^2 \tag{66}$$

The reader is reminded that in Eqs. (59)–(66) the ideal value of D_n is in all cases D_e. The intercepts I_n are also useful kinetic parameters. Their ideal values are

$$I_5^0 = I_7^0 = \ln (2/\pi^2), \quad I_2^0 = I_{2M}^0 = I_8^0 = \ln (8/\pi^2).$$

Considerable progress has recently been made in developing the theoretical background necessary for the application of the above method of transient kinetic analysis. An important step in this direction was the use of WKB asymptotics to derive approximate analytical expressions for short- and long-time transient sorption and permeation in membranes characterized by concentration-independent continuous S(X) and $D_T(X)$ functions [150–154]. The earlier papers dealing with this subject [152–154] are referred to in a recent review [9]. The more recent articles [150,151] provide the correct asymptotic expressions applicable to all kinetic regimes listed above; the usefulness of these expressions is evaluated on the basis of complete numerical solutions of carefully selected examples; and the relation between transient transport behaviour and the nature and degree of the nonhomogeneity of sorption and diffusion properties is investigated in some detail.

Considerable attention has also been given to laminates with concentration independent solubility and diffusion coefficients [155-167]. The main object of this work was to derive analytical transient state solutions for symmetrical and unsymmetrical sorption [155,158,159,162-164,166], as well as permeation [141,156,161,166,167], and to illustrate their practical use [141,155,162,163,165]. Attention has also been given to boundary conditions other than the standard ones considered above, notably to well-stirred finite-bath conditions leading to variation of the boundary activities in sorption or of the downstream boundary activity in permeation [155,156,159,161]. From the point of view of technical practice, such boundary conditions are not usual in membrane separations; but they may be important in other applications, notably packaging where laminates are in wide use. The most general problems considered so far include symmetrical sorption in ternary (ABC type) laminates [158,166] and unsymmetrical sorption in, or permeation through, N-fold laminates [159,161]. More general problems (including, for example, source or sink terms or interfacial resistances to flow at lamination junctions) and other types of boundary conditions have been dealt with in the heat transfer field [168].

The solutions referred to above are useful for calculating transient transport behaviour given the sorption and diffusion properties of each component, and have been successfully tested experimentally in a number of instances [155,162,163,165]. The inverse problem can, in principle, be handled, but the procedure may be impractical except in simple cases. Thus, for example, suitable iterative procedures are suggested for this purpose on the basis of long-time sorption kinetic data, provided that the composition of the laminate can be varied [158]. The problem is greatly simplified if there is only one unknown component [158,165]. Problems of this type can be important in practice if there is some component which cannot be studied in isolation (a stagnant layer of the external bath adhering to the surfaces of the membrane during a sorption experiment is a case in point).[158] In the case of periodic laminates (e.g. ABAB . . .) there is the added simplification that ideal behaviour is rapidly approached as the number of periods increases [162,164,165,169,170].

However, from the point of view of the fundamental understanding of transient transport behaviour as a function of the nature and degree of the nonhomogeneity of the sorption and diffusion properties of the composite membrane, the general analytical solutions are not very informative, because of their complicated non-explicit character. Some numerical examples, together with a more systematic examination of sorption kinetics in ABAB . . . type laminates, have been reported [160,161,163,164,166]; but a thorough-going investigation of the type carried out for membranes characterized by continuous $S(X)$ and $D_T(X)$ functions [150,151] is still lacking.

Among the main results of the theoretical studies reviewed above, several general relations between the quantities measured in sorption and permeation experiments include [144,150]

$$Q_t^a = Q_t^{d*}; \qquad Q_t^d = Q_t^{a*}; \qquad M_t^a = M_t^d \tag{67}$$

where M_t denotes amount of penetrant sorbed in a symmetrical sorption experiment. Other relations of this kind are [143,144,151,154]

$$Q^a(0, t) = Q^{d*}(l, t) ; \qquad Q^d(l, t) = Q^{a*}(0, t)$$

$$Q^a(l, t) = Q^{a*}(l, t) = Q^d(0, t) = Q^{d*}(0, t) \tag{68}$$

$$\Delta Q_t^a = \Delta Q_t^{d*}; \qquad \Delta Q_t^d = \Delta Q_t^{a*} \tag{69}$$

$$M_t = \Delta Q_t^a + \Delta Q_t^d = \Delta Q_t^a + \Delta Q_t^{a*} \tag{70}$$

In all late-time regimes notably those represented by Eqs. (60), (62), (64) and (66) ideal kinetics is obeyed [144,150,151,154,159,161]; whereas this is not so in the short-time regimes presented by Eqs. (59), (63) and (65) [150,151,154,163,164] (which convey essentially the same information).[151] The short time behaviour of lower permeation curves represented by Eq. (61) appears to occupy an intermediate position, in the sense that ideal kinetics appears to be followed only to a first approximation [151]. The relation between permeation and symmetrical sorption indicated by Eq. (70) is also notable. The respective kinetics become very similar at long times [154] as indicated by the relevant relations [151] $D_{2M} = D_5 = D_7 = D_8$ and

$$I_{2M} = \ln [(\Delta Q_\infty/Q_\infty) \exp I_8 + (\Delta Q_\infty^*/Q_\infty) \exp I_8^*] .$$

It has been shown [150,151] that transient kinetic behaviour, like time lags, essentially reflects the properties of the function

$$H(y) = D_T(X) S(X)^2/D_e S_e^2$$

where

$$y = (lS_e)^{-1} \int_0^X S(z) \, dz$$

The detailed study of Ref. [150,151] indicates that the information about $H(y)$ conveyed by different kinetic regimes is in part similar and in part different. The similarity between late time permeation and symmetrical sorption kinetics has already been pointed out above. Symmetrical sorption kinetics at both short and long times are shown to reflect primarily the properties of d^2H/dy^2; whereas short time unsymmetrical sorption is mainly sensitive to dH/dy. For more detailed information the original papers [150,151] should be consulted.

Some preliminary kinetic analyses of transient sorption and permeation data have been reported [144,149]. Examples from kinetic regimes where ideal kinetics is expected to be obeyed are shown in Fig. 17. The linearity of the relevant plots is satisfactory and the values of D_3, D_4 and D_2^a, D_2^d determined from these plots deviate from D_e in opposite directions, as expected [150,151]; even though the detailed spatial dependence of S and D_T in the membrane in question is two-dimensional [145] (cf. previous subsection) and is, therefore, more complicated than envisaged in the theoretical

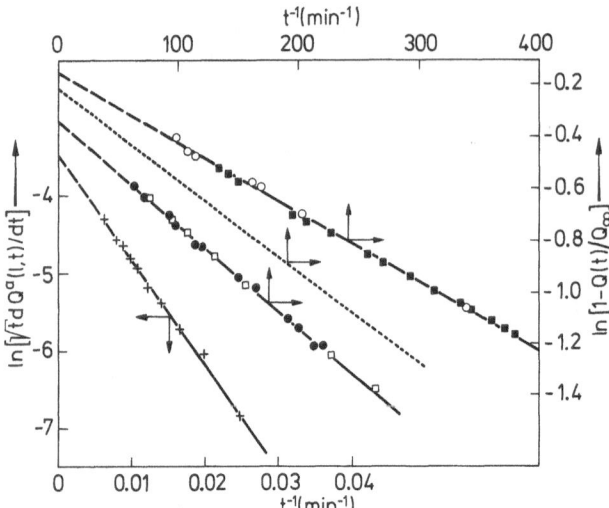

Fig. 17. Experimental long-time Q_t and short-time $Q^a(l, t)$ transient kinetic plots, according to Eqs. (60) and (61) respectively, for N_2 at 297 K in Membrane C of Table 5: $+ Q^a(l, t)$; $\bullet Q_t^a$; $\bigcirc Q_t^d$; $\blacksquare Q_t^{a*}$; $\square Q_t^{d*}$; ... corresponding ideal long-time Q_t plot extrapolated to $t = 0$. [144] Effective diffusion coefficients (in $cm^2 s^{-1} \times 10^{-4}$): $D_e = 0.30$; $D_3 = 0.39$, $D_4 = 0.20$; $D_2^a = 0.34$, $D_2^d = 0.23$. Intercepts: $I_2^a = -0.35$, $I_2^d = -0.15$ ($I_2^0 = -0.21$) [144]

studies reviewed above. Examples of analogous plots for a system where S and D_T are both spatially and concentration dependent have also been given [149].

Transient sorption and permeation data may also be examined in the form of suitable moments. For example, in the case of sorption one may define the following moments [171]

$$W_n = \int\limits_0^\infty t^n Q_t \, dt \qquad (n = 1, 2, ...)$$

These parameters are in many ways analogous to permeation time lags; but the relevant expressions for the case of S(X), $D_T(X)$ are considerably more complicated (even in the case of the first moment which is the simplest) [171] than the corresponding time lag formulae [4]. Accordingly, moments represent a less efficient way of making use of the information contained in transient diffusion data than the methods discussed above. In spite of these limitations, further study of moments should prove worthwhile.

The above discussion provides an illustration of the wealth of information that can be obtained from the transient transport behaviour of non-ideal penetrant-membrane systems. The methods discussed are capable of further development and refinement and are potentially applicable to a wide variety of experimental systems.

Acknowledgement: Valuable detailed comments from Profs. W. J. Koros and T. Iijima are gratefully acknowledged.

References

1. Petropoulos J. H., Roussis P. P.: J. Chem. Phys. *47*, 1491 (1967)
2. Crank, J., Park G. S. (eds.): Diffusion in Polymers, Academic Press, New York, 1968
3. Crank, J.: The Mathematics of Diffusion, 2nd Ed., Clarendon Press, Oxford, 1975
4. Petropoulos J. H., Roussis P. P.: J. Chem. Phys. *47*, 1496 (1967)
5. Petropoulos J. H., Roussis P. P.: J. Chem. Phys. *50*, 3951 (1969)
6. Vieth W. R., Howell J. M., Hsieh J. H.: J. Membrane Sci. *1*, 177 (1976)
7. Paul D. R.: Ber. Bunsenges. Phys. Chem. *83*, 294 (1979)
8. Stannett V. T., Koros W. J., Paul D. R., Lonsdale H. K., Baker R. W.: Adv. Polym. Sci. *32*, 69 (1979)
9. Frisch H. L., Stern S. A.: C.R.C. Crit Rev. Solid State Mater. Sci. *11*, 123 (1983)
10. Chern R. T., Koros W. J., Sanders E. S., Chen S. H., Hopfenberg H. B.: in Industrial Gas Separations, Chap. 3. Whyte T., Yon C., Wagener E. (ed.). Am. Chem. Soc. Symp. Ser. No 223 (1983)
11. Michaels A. S., Vieth W. R., Barrie J. A.: J. Appl. Phys. *34*, 1 (1963)
12. Koros W. J., Paul D. R.: J. Polym. Sci., Polym. Phys. Ed. *16*, 1947 (1978)
13. Koros W. J., Paul D. R.: Polym. Eng. Sci. *20*, 14 (1980)
14. Koros W. J., Paul D. R., Huvard G. S.: Polymer *20*, 956 (1979)
15. Huvard G. S., Stannett V. T., Koros W. J., Hopfenberg H. B.: J. Membrane Sci. *6*, 185 (1980)
16. Felder R. M., Patton C. J., Koros W. J.: J. Polym. Sci., Polym. Phys. Ed. *19*, 1895 (1981)
17. Koros W. J., Patton C. J., Felder R. M., Fincher S. J.: J. Polym. Sci., Polym. Phys. Ed. *18*, 1845 (1980)
18. Koros W. J., Smith G. N., Stannett V. T.: J. Appl. Polym. Sci., *26*, 159 (1981)
19. Chan A. H., Paul D. R.: J. Appl. Polym. Sci. *24*, 1539 (1979)
20. Wonders A. G., Paul D. R.: J. Membrane Sci., *5*, 63 (1979)
21. Yavorsky J. A., Spencer H. G.: J. Appl. Polym. Sci. *25*, 2109 (1980)
22. Chan A. H., Paul D. R.: Polym. Eng. Sci., *20*, 87 (1980)
23. Erb A. J., Paul D. R.: J. Membrane Sci., *8*, 11 (1981)
24. Chan A. H., Koros W. J., Paul D. R.: J. Membrane Sci. *3*, 117 (1978)
25. Toi K., Morel G., Paul D. R.: J. Appl. Polym. Sci. *27*, 2997 (1982)
26. Masi P., Paul D. R., Barlow J. W.: J. Polym. Sci., Polym. Phys. Ed. *20*, 15 (1982)
27. Barrie J. A., Williams M. J. L., Munday K.: Polym. Eng. Sci., *20*, 20 (1980)
28. Toi K., Paul D. R.: Macromolecules, *15*, 1104 (1982)
29. Toi K., Maeda Y., Tokuda T.: J. Membrane Sci., *13*, 15 (1983)
30. Stern S. A., De Meringo A. H.: J. Polym. Sci., Polym. Phys. Ed. *16*, 735 (1978)
31. Stern S. A., Kulkarni S. S.: J. Membrane Sci. *10*, 235 (1982)
32. Berens A. R.: Polym. Eng. Sci. *20*, 95 (1980)
33. Grzywna Z. J., Podkowka J.: J. Membrane Sci. *8*, 23 (1981)
34. Raucher D., Sefcik M. D.: in Industrial Gas Separations, Chap. 6. Whyte T., Yon C., Wagener E. (ed.). Am. Chem. Soc. Symp. Ser. No 223 (1983)
35. Barrer R. M.: Zeolites and Clay Minerals as Sorbents and Molecular Sieves., Chap. 3, 4. New York: Academic Press 1978
36. Young D. M., Crowell A. D.: Physical Adsorption of Gases. Chap. 4, 6. London: Butterworth 1962
37. Youngquist G. R., Allen J. L., Eisenberg J.: Ind. Eng. Chem. Prod. Res. Develop. *10*, 308 (1971)
38. Koros W. J., Paul D. R.: J. Polym. Sci., Polym. Phys. Ed. *19*, 1655 (1981)
39. Koros W. J., Chan A. H., Paul D. R.: J. Membrane Sci., *2*, 165 (1977)
40. Enscore D. J., Hopfenberg H. B., Stannett V. T.: Polymer *18*, 793 (1977)
41. Fechter J. M. H., Hopfenberg H. B., Koros W. J.: Polym. Eng. Sci. *21*, 925 (1981)
42. Michaels A. S., Baddour R. F., Bixler H. J., Choo C. Y.: Ind. Eng. Chem. Proc. Des. Develop. *1*, 14 (1962)
43. Sefcik M. D., Schaefer J.: J. Polym. Sci., Polym. Phys. Ed. *21*, 1055 (1983)
44. Berens A. R., Angew. Makromol. Chem. *47*, 97 (1975)
45. Stannett V., Haider M., Koros W. J., Hopfenberg H. B.: Polym. Eng. Sci. *20*, 300 (1980)
46. Ranade G., Stannett V., Koros W. J.: J. Appl. Polym. Sci. *25*, 2179 (1980)

47. Saxena V., Stern S. A.: J. Membrane Sci. *12*, 65 (1982)
48. Mauze G. R., Stern S. A.: J. Membrane Sci., *12*, 51 (1982)
49. Mauze G. R., Stern S. A.; The Dual-Mode Solution and Transport of Water in Poly (acryloni-
    trile), to be published
50. Sefcik M. D., Raucher D.: The Matrix Model of Gas Sorption and Diffusion in Glassy Polymers,
    to be published
51. Koros W. J.: personal communication
52. Petropoulos J. H.: J. Polym. Sci. *A2 8*, 1797 (1970)
53. Paul D. R., Koros W. J.: J. Polym. Sci., Polym. Phys. Ed. *14*, 675 (1976)
54. Koros W. J., Paul D. R., Rocha A. A.: J. Polym. Sci., Polym. Phys. Ed. *14*, 687 (1976)
55. Koros W. J., Paul D. R.: J. Polym. Sci., Polym. Phys. Ed. *16*, 2171 (1978)
56. Chan A. H., Paul D. R.: J. Appl. Polym. Sci. *25*, 971 (1980)
57. Toi K.: Polym. Eng. Sci. *20*, 30 (1980)
58. Toi K., Ohori Y., Maeda Y., Tokuda T.: J. Polym. Sci., Polym. Phys. Ed. *18*, 1621 (1980)
59. Frisch H. L.: J. Phys. Chem. *60*, 1177 (1956)
60. Stannett V. T.: Chap. 2 in Ref. 2
61. Koros W. J., Hopfenberg H. B., Petropoulos J. H.: unpublished
62. Vrentas J. S., Duda J. L.: J. Appl. Polym. Sci. *22*, 2325 (1978)
63. Raucher D., Sefcik M. D.: Polym. Prepr. Div. Polym. Chem., Am. Chem. Soc. *24*, 85 (1983)
64. Stern, S. A., Saxena V.: J. Membrane Sci. *7*, 47 (1980)
65. Morel, G., Paul, D. R.: J. Membrane Sci. *10*, 273 (1982)
66. Tanioka, A., Oobayashi, A., Kageyama, Y., Miyasaka, K., Ishikawa, K.: J. Polym. Sci., Polym.
    Phys. Ed. *20*, 2197 (1982)
67. Koros, W. J.: J. Polym. Sci., Polym. Phys. Ed. *18*, 982 (1980)
68. Koros, W. J., Chern, R. T., Stannett, V., Hopfenberg, H. B.: J. Polym. Sci., Polym. Phys Ed.
    19, 1513 (1981)
69. Sanders, E. S., Koros, W. J., Hopfenberg, H. B., Stannett, V. T.: J. Membrane Sci. *13*, 161
    (1983)
70. Sanders, E. S., Koros, W. J., Hopfenberg, H. B., Stannett, V. T.: Pure and Mixed Gas Sorption
    of Carbon Dioxide and Ethylene in Poly(methyl methacrylate), to be published
71. Chern, R. T., Koros, W. J., Hopfenberg, H. B., Stannett, V. T.: J. Polym. Sci., Polym. Phys.
    Ed. *21*, 753 (1983)
72. Chern, R. T., Koros, W. J., Sanders, E. S., Yui, R.: J. Membrane Sci. *15*, 157 (1983)
73. Chern, R. T., Koros, W. J., Yui, B., Hopfenberg, H. B., Stannett V. T.: Selective Permeation
    of CO_2 and CH_4 through Kapton Polyimide: Effects of Penetrant Competition and Gas Phase
    Non-ideality, to be published
74. Iijima, T., Kim, J.: Sen-i Gakkaishi, *37*, P353 (1981)
75. Vickerstaff, T.: The Physical Chemistry of Dyeing, 2nd Ed. London: Oliver and Boyd 1954
76. Peters, R. H.: The Physical Chemistry of Dyeing, Textile Chemistry, *Vol. 3*, New York: Elsevier
    1975
77. Atherton, E., Downey D. A., Peters R. H.: Textile Res. J. *25*, 977 (1955)
78. Palmer, H. J.: J. Textile Inst. *49*, T33 (1958)
79. Milicević, B., McGregor, R.: Textilveredlung *2*, 130 (1967)
80. Komiyama, J., Petropoulos, J. H., Iijima, T.: J. Soc. Dyers Colourists *93*, 217 (1977)
81. Komiyama, J., Iijima, T.: J. Polym. Sci., Polym. Phys. Ed. *12*, 1465 (1974)
82. Tak, T., Komiyama, J., Iijima, T.: Sen-i Gakkaishi *35*, T487 (1979)
83. Sasaki, S., Komiyama, J., Iijima, T.. Petropoulos, J. H.: unpublished
84. Tak, T., Sasaki, T., Komiyama, J., Iijima, T.: J. Appl. Polym. Sci. *26*, 3325 (1981)
85. Rice, R. G., Foo, S. C.: Ind. Eng. Chem. Fundam., *18*, 63 (1979)
86. Foo, S. C., Rice, R. G.: Ind. Eng. Chem. Fundam., *18*, 68 (1979)
87. Iijima, T., Miyata, E., Komiyama, J.: Polym. Eng. Sci. *20*, 271 (1980)
88. Barrer, R. M.: Chap. 6 in Ref [2]
89. See Chap. 12 in Ref [3]
90. Petropoulos, J. H.: Remarks on the Theoretical Description of the Permeability Properties of
    Binary Composite Polymers, Paper presented at the 4th I.U.P.A.C. Discussion Conference on
    Macromolecules, Marianske Lazne (Marienbad), Sept. 1974: to be published
91. De Vries, D. A.: Bull. Inst. Internat. Froid, Annexe 1952-1, p. 115

92. Barrer, R. M., Petropoulos, J. H.: Brit. J. Appl. Phys. *12*, 691 (1961)
93. Hopfenberg, H. B., Paul, D. R.: in Polymer Blends, Vol. I, Chap. 10. Paul, D. R., Newman, S. (ed.). New York: Academic Press 1978
94. Odani, H., Taira, K., Nemoto, N., Kurata, M.: Polym. Eng. Sci. *17*, 527 (1977)
95. Odani, H., Uchikura, M., Ogino, U., Kurata, M.: J. Membrane Sci., *15*, 193 (1983)
96. Odani, H., Uchikura, M., Taira, K., Kurata, M.: J. Macromol. Sci., Phys. *B17*, 337 (1980)
97. Chiang, K. T., Sefton, M. V.: J. Polym. Sci., Polym. Phys. Ed. *15*, 1927 (1977)
98. Ostler, M. I., Rogers, C. E.: Effects of Interphase Structure on Gas Sorption and Diffusion in Styrene-Butadiene Triblock Polymers, to be published
99. Yamada, S., Nakagawa, T.: Kobunshi Ronbunshu *34*, 683 (1977)
100. Higuchi, W. I.: J. Phys. Chem. *62*, 649 (1958)
101. Barrie, J. A., Munday, K.: J. Membrane Sci. *13*, 175 (1983)
102. Barrie, J. A., Ismail, J. B.: J. Membrane Sci. *13*, 197 (1983)
103. Barbier, J.: Rubber Chem. Tech. *31*, 814 (1954)
104. Barrie, J. A., Munday, K., Kuo, D., Ismail, J., Williams, N., Spencer, H. G.: Gas Transport in Graft and Block Copolymers, Paper presented at the I.U.P.A.C. 28th International Conference on Macromolecules, Amherst (Mass), July 1982
105. Paul, D. R.: Gas Transport in Homogeneous Multicomponent Polymers, J.Membrane Sci., in press
106. Tsimboukis, D. G., Petropoulos, J. H.: J. Chem. Soc. Faraday I *75*, 717 (1979)
107. McGregor, R., Harris, P. W.: J. Appl. Polym. Sci. *14*, 513 (1970)
108. Kobatake, Y., Kurihara, K., Kamo, N. in: Surface Electrochemistry, Chap. 1. Takamura, T., Kozawa, A. (ed.). Tokyo: Japan Scientific Societies Press 1978
109. Helfferich, F.: Ion Exchange, Chap. 5., New York: McGraw-Hill 1962
110. Tsimboukis, D. G., Petropoulos, J. H.: J. Chem. Soc. Faraday I *75*, 705 (1979)
111. Meares, P.: Chap. 10 in Ref. 2
112. Glueckauf, E.: Proc. Roy. Soc. *A 268*, 350 (1962)
113. Narębska, A., Wódzki, R.: Angew. Makromol. Chem. *80*, 105 (1979)
114. Schofield, R. K., Talibuddin, O.: Disc. Faraday Soc. *3*, 51 (1948)
115. Sivaraja-Iyer, S. R. and Jayaram, R.: J. Soc. Dyers and Colourists, *86*, 398 (1970)
116. Sivaraja-Iyer, S. R., in: The Chemistry of Synthetic Dyes, Vol. VII, Chap. IV, p. 201. New York: Academic Press 1974
117. Vink, H.: Acta Chem. Scand, *A33*, 547 (1979)
118. Thanh, V. V., Petropoulos, J. H., Iijima, T.: J. Appl. Polym. Sci. *28*, 2813 (1983)
119. Glueckauf, E., Watts, R. E.: Proc. Roy. Soc. *A268*, 339 (1962)
120. Wódzki, Narębska, A., Ceynowa, J.: Angew. Makromol. Chem. *78*, 145 (1979)
121. Petropoulos, J. H., Tsimboukis, D. G., Kouzeli, K.: J. Membrane Sci. *16*, 379 (1983)
122. Kokotov, Yu. A.: Russ. J. Phys. Chem. *52*, 94 (1978)
123. Kokotov, Yu. A.: Russ. J. Phys. Chem. *52*, 96 (1978)
124. Crabtree, J. M., Glueckauf, E.: Trans. Faraday Soc. *59*, 2639 (1963)
125. Gierke, T. D., Munn, G. E., Wilson, F. C.: J. Polym. Sci., Polym. Phys. Ed. *19*, 1687 (1981); and refs therein
126. Hsu, W. Y., Gierke, T. D.: J. Membrane Sci. *13*, 307 (1983)
127. Glueckauf, E.: Proc. First Int. Symp. Water Desalination, Washington D. C., Oct. 1965 *1*, 143 (1967)
128. Merten, U., in: Desalination by Reverse Osmosis, Chap. 2, Merten U. (ed.). Cambridge (Mass): M.I.T. Press 1966
129. Bean, C. P., in: Membranes, A Series of Advances, Vol. 1, Chap. 1. Eisenman G. (ed.). New York: Marcel Dekker 1972
130. Taniguchi, Y., Horigome, S.: Desalination *16*, 395 (1975)
131. e.g. Demisch, H. U., Pusch, W.: J. Electrochem. Soc. *123*, 370 (1975)
132. Thau, G., Bloch, R., Kedem, O.: Desalination *1*, 129 (1966)
133. Glueckauf, E., Russell, P. J.: Desalination *8*, 351 (1970)
134. Anderson, J. E., Pusch, W.: Ber Bunsengesell. Physik. Chem., *80*, 847 (1976)
135. Stannett, V., Hopfenberg, H. B., Petropoulos, J. H., in: Macromolecular Science, Chap. 8. Bawn C.E.H. (ed.). Vol. 8 in MTP International Review of Science. London: Butterworths 1972

136. Petropoulos, J. H.: J. Polym. Sci., Polym. Phys. Ed. *11*, 1867 (1973)
137. Petropoulos, J. H.: J. Polym. Sci., Polym. Phys. Ed. *12*, 35 (1974)
138. Rogers, C. E., Stannett, V., Szwarc, M.: Ind. Eng: Chem. *49*, 1933 (1957)
139. C. E. Rogers: J. Polym. Sci. *C10*, 93 (1965)
140. Penati, A., Pegoraro, M.: J. Appl. Polym. Sci. *22*, 3213 (1978)
141. Choji, M., Matsuura, I., Karasawa, M.: Sen-i Gakkaishi *37*, T192 (1981)
142. Roussis, P. P., Petropoulos, J. H.: J. Chem. Soc. Faraday II *72*, 737 (1976)
143. Roussis, P. P., Petropoulos, J. H.: J. Chem. Soc., Faraday II *73*, 1025 (1977)
144. Tsimillis, K., Petropoulos, J. H.: J. Phys. Chem. *81*, 2185 (1977)
145. Savvakis, C., Petropoulos, J. H.: J. Phys. Chem. *86*, 5128 (1982)
146. Petropoulos, J. H., Roussis, P. P.: J. Chem. Phys. *48*, 4619 (1968)
147. Ash, R., Barrer, R. M., Chio, H. T., Edge, A. V. J.: Proc. R. Soc. Lond. *A365*, 267 (1979)
148. Roussis, P. P., Tsimillis, K., Savvakis, C., Amarantos, S., Kanellopoulos, N., Petropoulos, J. H.: J. Phys. *E13*, 403 (1980)
149. Amarantos, S. G., Tsimillis, K., Savvakis, C., Petropoulos, J. H.: J. Membrane Sci. *13*, 259 (1983)
150. Grzywna, Z. J., Petropoulos, J. H.: J. Chem. Soc. Faraday II *79*, 571 (1983)
151. Grzywna, Z. J., Petropoulos, J. H.: J. Chem. Soc. Faraday II *79*, 585 (1983)
152. Frisch, H. L., Bdzil, J.: J. Chem. Phys. *62*, 4804 (1975)
153. Frisch, H. L.: J. Membrane Sci. *82*, 1559 (1978)
154. Frisch, H. L.: J. Phys. Chem. *83*, 149 (1978)
155. Spencer, H. G., Barrie, J. A.: J. Appl. Polym. Sci. *22*, 3539 (1978)
156. Spencer, H. G., Barrie, J. A.: J. Appl. Polym. Sci. *23*, 2537 (1979)
157. Spencer, H. G., Barrie, J. A.: J. Appl. Polym. Sci. *24*, 1391 (1979)
158. Spencer, H. G., Barrie, J. A.: J. Appl. Polym. Sci. *25*, 1157 (1980)
159. Spencer, H. G., Barrie, J. A.: J. Appl. Polym. Sci. *25*, 2807 (1980)
160. Spencer, H. G., Chen, T. C.: J. Appl. Polym. Sci. *26*, 2797 (1981)
161. Spencer, H. G., Chen, T. C., Barrie, J. A.: J. Appl. Polym. Sci. *27*, 3835 (1982)
162. Choji, N., Karasawa M.: Sen-i Gakkaishi *34*, T207 (1978)
163. Choji, N., Karasawa, M., Nagasawa, H., Matsuura, I.: Sen-i Gakkaishi *34*, T274 (1978)
164. Choji, N., Karasawa, M.: Sen-i Gakkaishi *36*, T451 (1980)
165. Choji, N., Karasawa, M., Nagasawa, H., Matsuura, I.: Sen-i Gakkaishi *36*, T459 (1980)
166. G. J. le Poidevin: J. Appl. Polym. Sci. *27*, 2901 (1982)
167. Berner, B., Cooper, E. R.: J. Membrane Sci. *14*, 139 (1983)
168. Wirth, P. E., Rodin, E. Y.: Adv. Heat Transf. *15*, 283 (1982)
169. Petropoulos, J. H., Roussis, P. P.: J. Chem. Phys. *51*, 1332 (1969)
170. Falkovitz, M. S., Frisch, H. L.: J. Membrane Sci. *10*, 61 (1982)
171. Frisch, H. L., Forgacs, G., Chui, S. T.: J. Phys. *83*, 2787 (1979)

M. Gordon (Editor)
Received October 10, 1983

Author Index Volumes 1–64

Subject Index

W. L. Hawkins

Polymer Degradation and Stabilization

Editor: H. J. Harwood

1984. 33 figures. XI, 119 pages
(Polymers – Properties and Applications, Volume 8)
ISBN 3-540-12851-4

Contents: Introduction. – Polymer Degradation. – Stabilization Against Non-oxidative Thermal Degradation. – Stabilization Against Thermal Oxidation. – Stabilization Against Degradation by Radiation. – Stabilization Against Degradation by Ozone. – Test Procedures. – Future Trends. – Subject Index.

The monograph is a concise review of the current status of research on those mechanisms responsible for the degradation of polymers when exposed to a hostile environment. Emphasis is placed on chemical reactions responsible for those degradation and stabilization processes for which there is a generally accepted mechanism. These include degradation that takes place on exposure to thermal, mechanical and radiation energy.

Mechanisms by which stabilizers inhibit degradation include those for autooxidation, ozone attack and photocoxidation. Stabilization by ultraviolet absorbers, radical traps, quenchers and hindered-amine-light stabilizers are described for a wide variety of polymers and the most recent research results are included. Procedures for testing the stability of polymers are reviewed. Both design and materials tests are evaluated. Problems encountered in extrapolating accelerated test data to actual use conditions are discussed.

This review will provide the background for selection of the most effective stabilizers or stabilizer combinations to protect a specific polymer against degradation. The subject is clearly and understandably presented. The book is of interest to undergraduate and graduate students, to industrial chemists and to chemists involved in polymer formulation.

Springer-Verlag
Berlin
Heidelberg
New York
Tokyo

Polymer Bulletin

Editors:

H.-J. Cantow, Freiburg; J. P. Kennedy, Akron, OH;
T. Saegusa, Kyoto

Editorial Board:

H. Batzer, Basel; S. Cesca, San Donato, Milanese; K. Dušek,
Prague; C. D. Eisenbach, Freiburg; P. J. Flory, Stanford, CA;
J. Furukawa, Tokyo; H. K. Hall, Jr., Tucson, AZ; M. L. Hal-
lensleben, Hannover; H. H. Kausch, Lausanne; T. Kelen,
Budapest; M. Kryszewski, Lódź; A. Ledwith, Liverpool;
R. W. Lenz, Amherst, MA; E. Maréchal, Paris; J. Meißner,
Zürich; A. Nakajima, Kyoto; E. F. Oleinik, Moscow; G. and
S. Henrici Olivé, Pensacola, FL; V. Percec, Cleveland, OH;
N. A. Platé, Moscow; C. I. Simionescu, Bucureşti; S. Sivaram,
Gujarat; D. H. Solomon, Melbourne; H. Tadokoro, Osaka;
M. Takayanagi, Fukuoka; I. Uematsu, Tokyo; O. Vogl,
Amherst, MA; C. Wippler, Strasbourg; H. Zahn, Aachen

Editorial Assistant:

A. Heinrich, Springer-Verlag Heidelberg

Character: betweeen the purely archival journals of full
papers and "letter journals" consisting exclusively of short
communications; length of papers, 4–8 pages

High-quality papers with an internationl spectrum

Competent referee system: high rejection rate

Rapid publication of papers: 3–6 weeks

50 reprints of each paper free of charge

No page charge

Springer
International

Subscription information and/or **sample** copies are available
from your bookseller or directly from Springer-Verlag,
Journal Promotion Dept., P. O. Box 105 280,
D-6900 Heidelberg, FRG

Orders from North America should be addressed to:
Springer-Verlag New York Inc., Journal Sales Department,
175 Fifth Ave., New York, 10010, USA

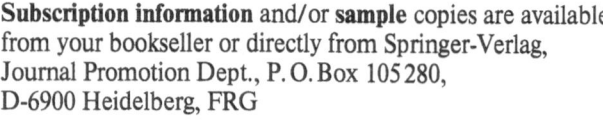